Innovating in Product/Process Development

Mikel Sorli • Dragan Stokic

Innovating in Product/Process Development

Gaining Pace in New Product Development

Springer

Mikel Sorli, Dr.
Fundación Labein – Tecnalia
Parque Tecnológico de Bizkaia
C/ Geldo
48160 Derio
Spain
sorli@labein.es

Dragan Stokic, Dr.
ATB Institut für Angewandte
Systemtechnik Bremen GmbH
Wiener Strasse 1
28359 Bremen
Germany
dragan@atb-bremen.de

ISBN 978-1-4471-5751-9 ISBN 978-1-84882-545-1 (eBook)
DOI 10.1007/978-1-84882-545-1
Springer Dordrecht Heidelberg London New York

British Library Cataloguing in Publication Data
A catalogue record for this book is available from the British Library

Cover design: eStudioCalamar, Figueres/Berlin

Printed on acid-free paper

Springer is part of Springer Science+Business Media (www.springer.com)

To our respective families.

"Design can be art. Design can be aesthetics.
Design is so simple, that's why it is so complicated"
Paul Rand
American Designer

"Design is not just what it looks like and feels like.
Design is how it works"
Jobs, Steven Paul
Co-founder & Chairman of Apple Computer Inc

Preface

The present book is intended to give an overview of the existing methods for product/process design and development and provoke discussion on the achievements and new trends for the twenty-first century including in the new proposed processes the relevant concept of innovation.

Innovation is a critical factor in the success of industrial companies, and just as important is the need to get innovative products to the market quickly. Therefore, it is important to talk about "management of product development time" because, under this new paradigm, companies capable of "mastering" the development time will launch the product into the market just spending the planned time and resources and at the moment when it will achieve higher acceptance ratios in the market. This will give back to the company a higher market share and faster market penetration.

The main objective should be to provide the means for stimulating the creation of innovative ideas in general, and specifically on potential product/process improvements and problem solving. These ideas have necessarily to be collected throughout the extended enterprise from people involved with the products and processes and should be developed into innovations in a project basis process. This in turn requires effective utilization of information and communication technologies (ICT).

The baseline of the book is product/process development and innovation in manufacturing industry, but most of the presented methods are applicable in a wide variety of industrial companies in different sectors. The concept of a product is considered in a broader sense, *i.e.,* it includes material products but also ICT products and services in general.

The book explores different aspects related to innovation processes in industry acting in the global economy in the twenty-first century and presents in detail several approaches to support this processes by ICT based knowledge management systems and collaborative working environments.

It has resulted from the authors working experience mainly in advanced research projects and has been conceived as a text book that may support students, practitioners, design engineers, and scientific people in general in their efforts to improve conjoint product/process designs and development with the overall objectives of achieving better innovative and sustainable products in shorter times. The book includes descriptions of many practical applications of the presented approaches which have been investigated within these projects. Since one of the focuses of the book is ICT support of the innovation process in industry, it also includes many references to advanced and commercially available ICT tools. However, the objective of the book is not to recommend specific tools and, therefore, the references to different ICT tools have only the purpose to provide an insight into some characteristic (but not necessarily the best available) examples of such ICT solutions.

The book is structured in six chapters: Chaps. 1 to 3 deal with methodological and conceptual aspects of design and development (historical background, innovation and new proposed methods); Chaps. 4 and 5 analyze ICT tools related to the subject (ICT tools supporting the development process and collaborative work); Chap. 6 discusses the new trends end emerging disciplines.

Acknowledgements

Authors are willing and happy to express their acknowledgement to the co-partners in the several European research projects mentioned throughout the text as well as to the European Commission[1] that has supported those and the responsible Project Officers for their valuable comments and feedback:

- AIM (IST-2001-52222): Acceleration of Innovative Ideas to Market
- ASSIST (COOP-CT-2004-512841): Knowledge-based Intelligent Design Assistant
- CuteLoop (ICT-2007216420): Customer in the Loop: Using Networked Devices Enabled Intelligence for Proactive Customers Integration as Drivers of Integrated Enterprise
- e-Mult (IST-2004-027212): European Multi-threaded Dynamic SME Networks for Market-Driven End-of-Life-Vehicles (ELV) Recycling
- InAmI (NMP-IST- 2004-016788): Innovative Ambient Intelligence Based Services to Support Life Cycle Management of Flexible Assembly and Manufacturing Systems
- K-NET (ICT-2007-215584): Services for Context Sensitive Enhancing of Knowledge in Networked Enterprises
- Know-Construct (COLL-CT-2004-500276):Internet Platform for Knowledge-based Customer Needs Management and Collaboration among Small and Medium sized Enterprises (SME) in Construction Industry
- LeanPPD (NMP2-LA 2008-214090) Lean Product/Process Development
- WECIDM (Asia IT&C ASI/B7-301/3152-99/72553) Web-enabled Collaboration in Intelligence Design and Manufacture

Another important group of people to be warmly thanked for their invaluable support and help, are the management and co-workers (current and former) in the institutes where the authors are employed.

Last but not least, we have to acknowledge our respective families for the time stolen from leisure time in "normal" life during the preparation of the book.

---

[1] Disclaimer: the material in this book related to these projects does not represent the opinion of the European Community, and the European Community is not responsible for any use that might be made of its content

Contents

Abbreviations

3GPP	3rd Generation Partnership Project
A&R	Automation & Robotics
AD	Axiomatic Design
AFD	Anticipatory Failure Determination
AmI	Ambience Intelligence
APDL	Axiomatic Product Development Lifecycle
API	Application Programming Interfaces
AS	Application Software
BPEL	Business Process Execution Language
CA	Customer Attributes
CAD	Computer Aided Design
CAI	Computer Aided Innovation
CAM	Computer Aided Manufacturing
CBR	Case-Based Reasoning
CCS	Core Collaborative Services
CE	Concurrent Engineering
CKB	Common Knowledge Base
CM	Context Modeling
CNC	Computer Numeric Control
CTC	Components Test Cases
CWE	Collaborative Working Environments
DD	Dynamic Database
DE	Directed Evolution
DFA	Design for Assembly
DFD	Design for Disassembly
DFM	Design for Manufacturing
DFMt	Design for Maintenance
DFx	Design for "x": anything relevant to design
DMAIC	Define – Measure – Analyze – Improve – Control
DOE	Design of Experiments
DP	Design Parameters
DSS	Design Support System
EBOK	Engineering Book of Knowledge
ebXML	e-business XML

EC	European Commission
e-CNM	Electronic Customer Needs Management
EDM	Engineering Data Management
EE	Extended Enterprise
e.g.	*exempli gratia* = for instance
e-KCS	Electronic Knowledge Community Support
ERP	Enterprise Resources Planning
EU	European Union
FAST	Function Analysis System Tree
FMEA	Failure Mode and Effect Analysis
FR	Functional Requirements
FTC	Functional Test Cases
GPS	Global Positioning System
GUI	Graphic User Interface
http	Hypertext Transfer Protocol
IAA	Insurance Application Architecture
IC	Input Constraints
ICT	Information and Communication Technologies
i.e.	*id est* = that is to say
IEM	Integrated Enterprise Modelling
IP	Internet Protocol
IPR	Intellectual Property Rights
IPS	Innovative Problem Solving.
ISO	International Organization for Standardization
IT	Information Technologies
JADE	Java Agent DEvelopment framework
JSQC	Japanese Society for Quality Control
KBE	Knowledge-based Engineering
KF	Knowledge Forum
KM	Knowledge Management
LDAP	Lightweight Directory Access Protocol
LSA	Latent Semantic Analysis
LVT	Logo Visual Technology
MES	Manufacturing Execution System
MILS	Multiple Independent Levels of Security
MIMOSA	Machinery Information Management Open Systems Alliance
MM	"Molecules" of Meaning
MSI	Management of Social Interactions
NTBF	New Technology-Based Firm
O&M	Operations and Maintenance
OASIS	Organization for the Advancement of Structured Information Standards
ODP	Open Distributed Processing
OECD	Organisation for Economic Co-operation and Development

OEE	Overall Equipment Efficiency
OEM	Original Equipment Manufacturer
OKP	One of a Kind Production
OMA	Object Management Architecture
OPAL	Object, Process, and Actor modelling Language
OSA-EAI	Open System Architecture for Enterprise Application Integration
OWL	Web Ontology language
PAM	Pluggable Authentication Modules
PBCE	Point-based Concurrent Engineering
PDA	Personal Digital Assistant
PDCA	Plan-Do-Check-Act
PDM	Product Data Management
PDPC	Process Decision Programme Chart
PLM	Product Life cycle Management
PV	Process Variables
QCC	Quality Costs Control
QFD	Quality Function Deployment
RA	Reference Architecture
RAM	Reliability, Availability, Maintainability Techniques
RBR	Rule-Based Reasoning
RDF	Resource Description Framework
ROI	Return on Investment
RSII	Regional Summary Innovation Index
RTD	Research and Development
S&T	Scientific & Technological
SBCE	Set-based Concurrent Engineering
SC	System Components
SCM	Supply Chain Management
SME	Small and Medium sized Enterprises
SMED	Single Minute Exchange of Die
SMIL	Synchronized Multimedia Integration Language
SMS	Short Message Services
SOA	Service Oriented Architecture
SOAP	Simple Object Access Protocol
SPC	Statistical Process Control
SW	Software
TCP	Transmission Control Protocol
TOC	Theory of Constraints
TPS	Toyota Production System
TQM	Total Quality Management
TRIZ	Theory for Innovative Problem Solving (Russian acronym)
UDDI	Universal Description, Discovery and Integration
UI	User Interface

UN/CEFACT	United Nations Centre for Trade Facilitation and Electronic Business
UN/EDIFACT	United Nations Electronic Data Interchange For Administration, Commerce and Transport
UNSPSC	United Nations Standard Product and Services
URI	Uniform Resource Identifier
VA	Value Analysis
VBE	Virtual Breeding Environment
VBN	Virtual Business Network
VE	Value Engineering
VM	Value Management
VPN	Virtual Private Network
VR	Virtual Reality
VTE	Virtual Testing Environment
W3C	World Wide Web Consortium
WSDL	Web Service Description Language
WS-I	Web Services Interoperability Organization
XML	eXtensible Markup Language

Chapter 1
Product/Process Development

Abstract Chapter 1 goes through the evolution of industry with time, paying special attention to four critical aspects: customer, quality, lead time and production size. This analysis shows the amazing fact that starting from craftsmanship the changes of paradigm with time have made a comeback to the same concept of "artistry" wrapped up with many new concepts, methodologies and tools that have arisen during more than a century of time that has elapsed since the industrial revolution in the nineteenth century to the current situation in the twenty-first century. The chapter also realizes an overview of the state of the art in product/process development and finally analyzes and describes a series of tools that should be used in this process. In summary, this chapter can be considered as an analysis of the state of the art that will set the basis for the new advanced proposals that are developed in later chapters.

1.1 History of Industrial Evolution

Making a brief analysis of the evolution of industrial activities since the time when the word *industry* actually came into use, until the present day, it is peculiar to observe from the modern standpoint that the evolution has followed a loop coming back to the original starting point.

We shall begin by referring to the commonly agreed most relevant characteristics of the current situation:

- *Global market.* Captive markets or brand loyalty does not exist any more; competition is no longer limited to the immediate geographical environment, but can come from anywhere in the world. Consequently, the best also has to be the best world-wide what is usually denominated "World Class Manufacturers".
- *High technological complexity of goods and services.* The process of conceiving, designing, manufacturing and marketing a product is so complicated that it is beyond the control of a single organization. As a result, *the management of*

the production chain takes on a great degree of complexity which is difficult to control without information and communication technologies (ICT) support.

- *Higher people demands.* The increase in the general level of culture of modern-day society has a twofold effect:
  - On the one hand, the consumer of our products is becoming increasingly more demanding (and has a greater availability of mechanisms for consulting, claiming, and getting help), which means that companies must make a greater effort to keep them satisfied and happy with their services. It is well known that a lost customer will prove to be very costly for a company, and triggers a chain of desertions.
  - Moreover, at in-house level, company employees and collaborators have a growing need to feel part of the company, motivated to work, and informed of the consequences of their work (among other aspects).
- *Legal regulations.* Not only must the manufacturer comply with product specifications or market requirements, but also an increasingly larger number of legal conditioning factors, safety requirements, or environmental legislation are becoming more and more numerous and compulsory.

Having defined the current situation, we shall now see how industry has evolved, focusing on four aspects, as shown in Table 1.1, which are *customer, quality, lead time, and production volume.*

Table 1.1 Historical industrial evolution

Phases	Customer	Quality	Lead time	Production volume
Craftsmanship	Quite important	Total quality (TQ)	Fairly long	Unitary
First Industrial Revolution	Almost unimportant	Quality control (QC)	Very long	Mass production
Second Industrial Revolution	Very low importance	Process control (PC)	Long	Mass production in shorter series
Third Industrial Revolution	Highest importance	Total quality management (TQM)	Quite short	"Lean production": very small series tending to unitary[1]

[1] The term "lean" production refers only to the production size in small batches *vs.* "mass" production in huge series. "Lean" production has a quite different meaning to "lean manufacturing" that will be discussed in Chap. 6 within lean design.

a) Craftsmanship. The era we shall call craftsmanship at the dawn of industrial civilization (which could practically be termed industrial prehistory) was characterized by:

- *The customer* was of the utmost importance due to his closeness to the craftsman and to what we might call the *interactive design* that developed between both of them. In many cases, moreover, the actual craftsman was the end-user of the product.
- *Quality.* Using modern terminology, we can talk of total quality (TQ), since the craftsman totally masters all the factors that intervene in the process: raw material, technology, manufacturing resources, *etc.* and has therefore mastered quality and can work to what nowadays is known as *total quality.*
- *Lead time.* The time factor (lead time) was not at all relevant at that time and may generally be regarded as *quite long.*
- *Production volume.* Unitary production. The concept of *mass production* is completely unknown as yet.

b) First Industrial Revolution. The turn of the twentieth century witnessed certain significant events that marked the subsequent industrial evolution. The concept "industry" really only makes sense as of this time. This great revolution came about with the appearance of the theories of organization and rationalization of the work (Taylor 1911; Gilbreth 1911; Gilbreth and Gilbreth 1917) which is based, in turn, on the division of labour ideas by their great forerunner Adam Smith (Smith 1998), the development of industrial statistics – which came somewhat later with Shewhart (1939) and others – and particularly the practical application of these theories with an innovative spirit and with great success by Henry Ford I in the manufacture of his ultra famous Ford T.

In this new context the four factors under analysis evolve to:

- *Customer.* The customer completely loses his/her importance since we are in a demand-driven market and everything that is made is easily sold. By this time, there is the well-known saying by Ford: "Anyone can buy the color of car they want, as long as it is black."
- *Quality.* Its only interest lies in inspecting the product to try to detect faults and correct them. Material is not very expensive and labour is really cheap. We are in a phase of pure inspection: quality control (QC).
- *Lead time.* The time variable becomes completely negligible, since products have a very long life (production for years) and there is no need to renew them in the short term.
- *Production Volume.* Production volume is massive (*mass production*). Mass production involves the production of large amounts of the same product over long periods.

c) Second Industrial Revolution. This refers to the set of subtle but significant changes that took place in industry at the end of the 1970s and the beginning of the 1980s. The driving force behind these changes is, without a shadow of a doubt, the mainstreaming of micro-computing and its massive introduction into industry, both in management processes and production systems. The most evident consequence of all this is the enormous possibility of generating and managing information that it offers us and, in terms of processes, greater monitoring and control possibilities. The analysis of the evolution of the four factors says that:

- *Customer.* The market dynamics begins to change and is transformed into a supply-driven market. The customers begin to enjoy a certain importance but for the moment nobody is too worried about them.
- *Quality.* Microcomputing allows a great leap forward and is beginning to *control the process*. The process will guarantee that parts are manufactured to the quality required.
- *Lead time.* It begins to get shorter. Competition increases and this calls for greater product renovation, and development times must subsequently be reduced.
- *Production volume.* The size of production batches begins to fall slowly. Making large series of one product along time no longer makes sense, as the product easily becomes obsolete and stocked production must come up to scratch.

d) Third Industrial Revolution. This term may be used to define the current situation which we described at the beginning; we are going to analyze the four factors:

- *Customer.* In the current market situation, the customer is the fundamental item and become the *King*, the target of the whole company efforts. This idea is the core of all the philosophy that underpins concepts as total quality management (TQM) and Quality Function Deployment (QFD) to be discussed later.

Quality. There has been a return to the original concept of *total mastery* of all the factors that affect the finished product by the manufacturer (craftsman). However, as the organization is now a large one, as opposed to the primitive individual or family craftsman, the concept of *Management* is brought in. A single person cannot *master quality* so quality has to be managed — what becomes TQM approach.

- *Lead time.* At the moment, time is the competitive factor par excellence. Markets are renewed very quickly and as a result development time (*lead time*) has to become very short.
- *Production volume.* The trend in reducing manufacturing batches has heightened and the tendency is towards shorter runs, almost unit runs in theory: Just-in-Time (JIT), unit batches, *etc.*

Summarising, the cycle has made a closed loop and industry has returned to its beginnings while maintaining the benefits and advancements achieved through this evolution. Fundamental concepts of current situation are:

- *Utter importance of the customer*
- *Product development time*
- *Trend towards unit production*
- *Total quality management (TQM)*

Some authors such as McDonough and Braungart in their book "Cradle to Cradle" (McDonough and Braungart 2002) start to mention a new industrial revolution based on the idea of *sustainability*. This interesting topic will be discussed later on (Chap. 6) when analyzing future trends.

1.2 Overview of Current Situation

It can be agreed that the current situation may also be characterized by the analysis of the status of the four above mentioned factors (*customer, quality, lead time* and *production volume*) that are going to be discussed in the following points plus some new ones that will be introduced further on in the book.

1.2.1 Utter Importance of the Customer

Confluence and alignment of customer's desires, company business strategy, and profitability are keys to the product's success as Fig. 1.1 shows.

Fig. 1.1 Key to product success

Quality Function Deployment (QFD) will be discussed later in different sections of the book as a very useful tool to achieve this alignment of customer's requirements and business strategy. For sure, the third parameter (profitability) is a result of good alignment of the latter.

So, generally speaking it can be said that product design has to be *driven by the customer* within the limiting frame of the overall business strategy.

1.2.2 Product Development Time

Company's capability to manage and "master" the product/process development cycle is the most important factor that conditions the ability of the product to pay back benefits to the company. Figure 1.2 and the associated explanation shows the rationale for this assertion.

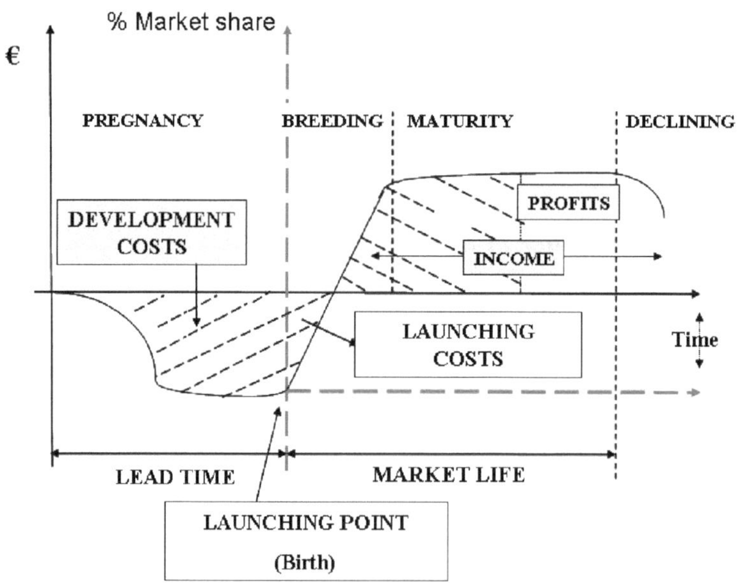

Fig. 1.2 Product development time

As may be seen in Fig. 1.2, the life span of any product can be divided into two big blocks:

- *Lead time.* Breeding from the idea generation to its conversion into a physical marketable product
- *Market life.* Following the "S" curve (*childhood, maturity, and declining to dead*)

Taking into account the economical factor, the *lead time* phase represents the inversion and expenditure encompassing development and launching costs while *market life* generates returns but the product becomes only profitable once the break even point is reached in which the previous inversion and costs have been recovered from sales income.

Figure 1.2 has two overlapped axes: the one in continuous line represents time in the x axis vs. money in the y axis (costs below the "0" and income above it). Coordinates in dotted lines show the product life cycle curve, expanding the well know *S curve* by addition of the pre-birth phase before the "0" time. The horizontal axis is again *time* and vertical represents *market penetration* (market share). For sure, the *market share* of the unborn product on the left side of the y axis, hasn't any meaning.

In the current situation *market life* is becoming quite short and replacement rate in many high technology edge products is actually in a single digit's year time. It is then mainly beyond the control of the manufacturing company since it is very much influenced by the evolution of the market, new developments by the competition, and the appearance of substitutive products.

In consequence, the company has to focus on mastering *lead time* where it has much more control. The usual approach of reducing lead time (and costs) is a very important issue but adjusting the launch date to the best possible time in terms of opportunity[2] is another relevant one to be considered.

Marketing campaigns and modern techniques to forecast market trends should help decision makers find the best launching date in which the maximum and quickest penetration of the product into the market may be achieved.

This will increase the gradient of the "launching" section of the "S-curve", reducing time needed (*childhood*) to achieve maturity on the market and also increase the market share percentage.

1.2.3 Trend Towards Unit Production

Craft times when every product was designed and produced upon a specific requirement from the customer, negotiated and developed together with the craftsman (the expert), have come nowadays to a situation in which every product is customized and adapted to the customer's needs.

In former times, each product was just "similar" to the previous one due to the fact that the basic design was coming out of the same brain and the product itself out of the same hands, but the standardization concept (basis of the serialization and mass production) was nonexistent.

---

[2] Hitting the best possible time to launch the product implies a quicker and higher market penetration curve.

First and second industrial revolutions (as discussed before) have come through standardization and mass production to stay. Their big advantages in costs make impracticable a comeback to the original craft situation but modern trends oblige us to adopt changes and solutions to emulate the old desirable situation by utilizing techniques looking for the best possible balance between scale economies and product customization.

The main approaches in that direction are:

- *In order to come closer to the customer needs*
  - *From the customer's point of view.* Widening the range of choices for the user to set up a "unique" product configuration: options, colors, variants, *etc.*
  - *From the manufacturing perspective.* Product modularization shifting from product "mass production" to modules "large production series" and final "customized" product assembly

- *In order to optimize production flows.* Lean manufacturing based on the so-called "Toyota Production System" (TPS) (Ashburn 1977; Monden 1987; Ohno 1988) is the key word. It can be defined as "Producing just what is needed, at the moment when is needed and at the point where is going to be used." A broad list of techniques has been developed under this umbrella:
  - JIT: just in time delivery
  - Kanban: shop floor material reposition
  - Poka-yoke: fools proof techniques
  - SMED: rapid tools and dies change (single minute exchange of die)
  - 6 σ: continuous improvement by using statistical techniques
  - Kaizen: continuous improvement by team working
  - Seven modes of waste: reduction of non-added value activities

1.2.4 Total Quality Management

Many attempts have been made to define total quality management (TQM) (Mizuno 1988; Ishikawa 1985). Resuming different contributions, it can be said that TQM is a philosophy and an ensemble of guiding principles that form the cornerstone of an organization that pursues continuous improvement. TQM entails the application of operating methods and human resources to improve goods, services, processes, *etc.* and, in a word, the degree to which customer needs are satisfied, now and in the future. TQM integrates management techniques, existing improvement efforts and technical tools, all under a disciplined approach targeting ongoing improvement. It is an attitude of every person in the organization, an incessant and permanent effort to improve in understanding, satisfying, and even overcoming customer expectations.

Total quality management, in a nutshell, is based on a cultural change in the company with the following basic pillars:

- *Customer satisfaction.* All company efforts should target the accomplishment of the objective of achieving maximum customer satisfaction through the product or service offered.
- *Deployment of policies.* Business objectives are deployed and itemized at different organizational levels, establishing partial consensus-based objectives among those in charge.
- *Involvement of all personnel.* Team work, existence of a true company culture, contribution inputs of all the personnel in the common goals, autonomy and promotion of each individual (empowerment), ongoing training, *etc.*
- *Ongoing improvement.* Continuous use of the well-known "plan-do-check-act" (PDCA) cycle[3] (Shewhart 1939; Deming 1986) at all levels, always pursuing ongoing improvement in all tasks at individual and team level.
- *Management by data.* Management of the necessary (and only necessary) information to monitor manufacturing and administrative processes at all times.
- *Management by processes.* Transformation of the traditional functional structures into new organizational forms targeting company processes, the most important ones being those that affect the external customer (customer satisfaction). One of the most important consequences of this approach is the trend towards much flatter organizations as opposed to the traditional hierarchical pyramids.

TQM entails two main aspects:

- All the company resources must be geared towards satisfying its customers' needs
- All the company employees and departments should be integrated in the Deming's cycle "plan-do-check-act" (PDCA) or cycle for continuous improvement (Deming 1986)

Simultaneous action in three areas is required (Akao 1991) for its application:

- *Vertical alignment.* This involves top management first of all which, by applying the "plan-do-check-act" (PDCA) cycle, states common objectives in a master plan targeting the customer, and transmits them to the whole organization from top to bottom to achieve alignment with them. This entails the existence of a company vision, strong leadership and the suitable use of different methods such as Policy Deployment, Benchmarking, the seven New Quality Management and Planning Tools, just to name some of them.

---

[3] PDCA cycle was first published by Shewhart but widely popularized later by Deming what has finally made it more known as Deming's PDCA cycle.

- *Horizontal integration.* Horizontal integration of the middle management levels, who participate and promote the deployment of policies, apply the PDCA cycle and engage in inter-functional management integrating the specific visions of each department or function. This involves a great deal of interfunctional team work, with the application of troubleshooting techniques, basic tools and new quality control and management tools: Statistical Process Control (SPC), Robust Design, JIT, QFD, *etc.* (see Sect. 1.4 and 1.5 for tools description).
- *Improvement of each unit.* Each person and department in the company should be involved in the ongoing improvement of their work processes by means of the application of the PDCA cycle and basic quality tools, maintaining and improving the quality level of their tasks.

The three aforementioned approaches complement rather than exclude each other, and act simultaneously in the TQM model, as represented in Fig. 1.3.

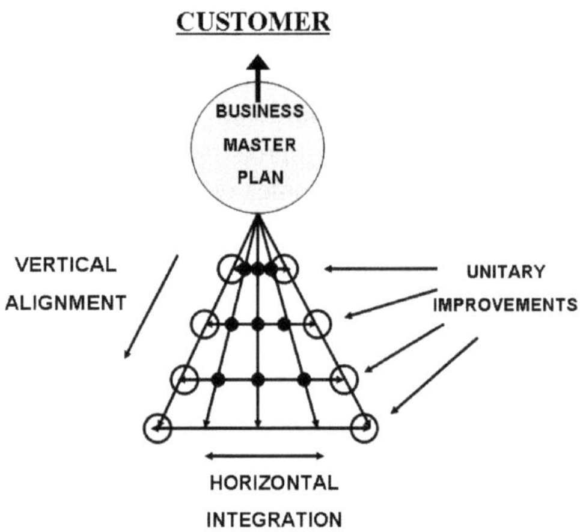

Fig. 1.3 Three vectors of total quality management (TQM)

Customer Oriented Master Plan. At the top of the pyramid is the *Master Plan* that must be "customer-oriented" collecting the approach of the organization over a 5–10 year time frame, based on its customers' needs. It envisions how the company will be transformed in this time frame to become the market leader in its specialty. It is oriented towards products and services, organizational effectiveness and, of course, profit. In a few words, it is the Strategic Plan of the company based on the customers/market.

The master plan (customer oriented) involves each and every one of the individuals of the organization, and begins with the identification of customer needs. It underlines the understanding of these needs and ongoing improvement in satisfying them.

1.3 Main Development Process

Product/process development process will be discussed and described in detail in Chap. 3 of this book in which a new process will be presented, integrating all modern trends, tools, and techniques.

The present section will deal with the big blocks of the development process together with the problems and drawbacks that the current situation entails.

It is not easy to say which can be considered old or new practices and when and where new proposals have been elaborated and applied. In general this distinction depends mainly on the different types of industrial sectors and varies with different geographical areas.

Generalizing, it can be said that product development has the phases shown in Fig. 1.4 (Pahl and Beitz 1996) and described in the following text. Vertical arrows in the figure show the different company departments involved in each phase. It may be seen that some overlapping exists to a small extent.

Phases. The following phases may be identified in every product development process:

1. *Start up decision.* Usually coming from top management with (desirable) participation of marketing.
2. *Specification definition.* Led by the design department, should be done in close relationship with the customer (if a direct customer) and participation of marketing (for consumer products).
3. *Conceptual design.* In which the overall characteristics and lay-out of the product will be defined.
4. *Detail design.* In which the conceptual design is broken down to detail drawings, material specifications, and production plans.
5. *Trials.* Production enters into the game and together with the design specialists develops the first physical products to check final appearance, feasibility, problem solving and other issues.
6. *Pre-production.* Short series are produced by the standard manufacturing means for the final fitness of machines, production systems, jigs and tools.
7. *Product launching.* The product is delivered to the unique customer or distributed to the sales net. Start of mass production.
8. *Market.* Marketing closes the loop checking acceptation of the product in the market, analysing evolution and providing feedback for new product launching.

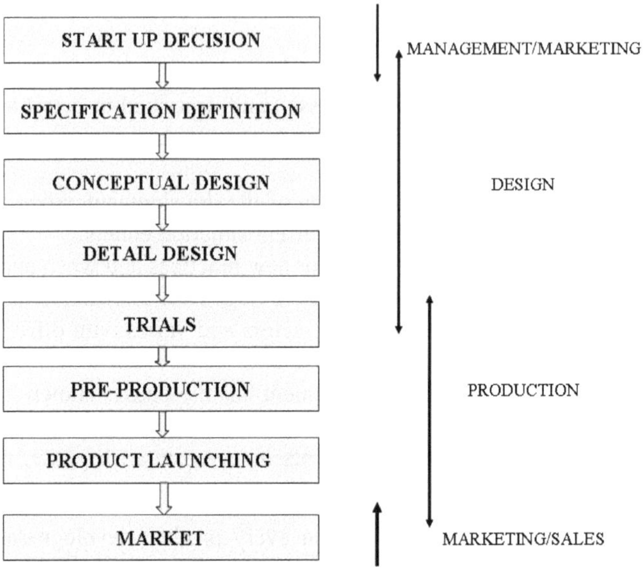

Fig. 1.4 Phases of product development

Though it is not mentioned, between each phase there is the need for a "*Design Review*" or methodical check out to assure that each next phase understands and translates correctly the requirements from the previous one. If there is any failure or missing information to be completed or reviewed, a return is required in order to correct or update the information.

This is one of the contributions from the quality system that is really more effective and necessary in order to make sure that the customer's requirements are correctly translated throughout the process to the final product.

Problems and drawbacks. Design process in traditional hierarchical organizations suffers a series of problems and drawbacks derived from the characteristics of working by functional departments, reporting through the hierarchy pyramid, *etc.* which can be summarized as:

- The process follows the design phases in a sequential way.
- Each department works in an isolated way with very low communication with other areas.
- Each department/person has knowledge on a very limited portion both of the overall product to be developed and of the whole process.
- Nobody in the organization has the whole overview of the picture. Top management has a plan based on figures, dates, and a basic idea of how the product will look.
- Traditional hierarchic structure generates complex and slow decision making processes.

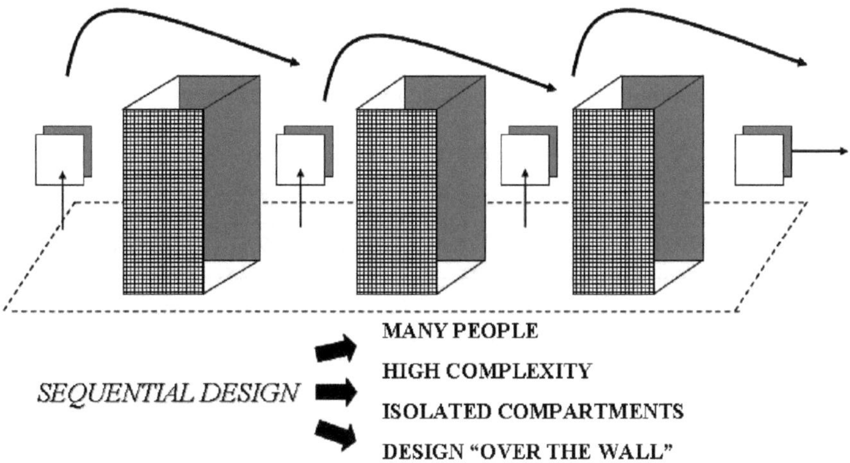

Fig. 1.5 Design over the wall

It can be said basically that the process is mostly sequential with very little overlapping and each functional department starts working once it receives the whole set of information and material from the previous section in the organization. This is widely known as "design over the wall," as Fig. 1.5 graphically shows, meaning that each department/section within the organization works in an isolated manner and the transmission of information is done "over the wall" almost without any human interaction.

Information flow and decision making processes are quite complex. Both have to pass through several levels and filters, making the decision making process very slow. This situation demonstrates that the decision is made at the top levels which are furthest away from the physical process and, most likely, not the best informed. The information has come to them through several filters, which in many cases may have adulterated it, and uses a shop floor language different from the one top management is used to.

Combination of all these factors makes the design process long, failure prone, and uncertain in results. As it is shown in Fig. 1.6, the length of the process comes not only from the addition of the duration of all phases but also from the idle time caused by delays in decision making (represented by the shaded small boxes).

Summarizing, it can be said that the design process in traditional organizations, is characterized by:

- *Designer.* He/she tends to do a convenient and closed design under time pressure. He/she has little time, and no motivation to innovate or find new approaches. The design has to be done quickly with the available information and there is neither time nor willingness to care about small details which may be the cornerstone of market success or the "spanner in the works" of the production line. Furthermore "Design for x" (DFx)[4] techniques are rarely taken into account and also, unfortunately, the designer rarely benefits from previous knowledge and in many cases is obliged to "reinvent the wheel."

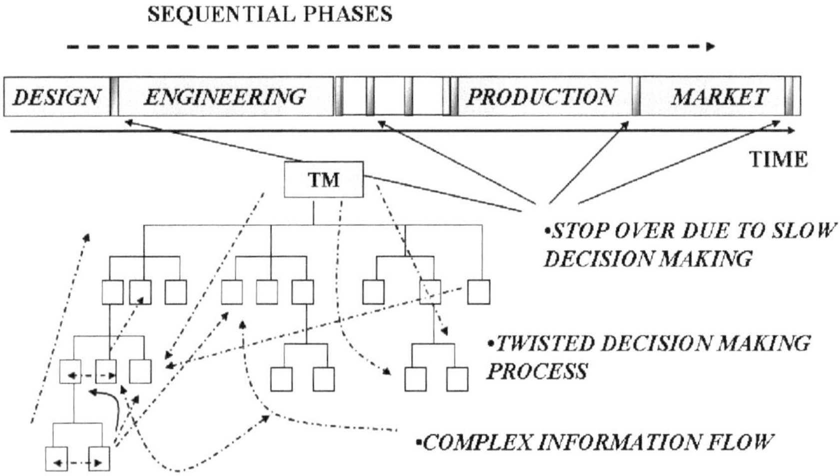

Fig. 1.6 Design process in traditional organizations

- *Paperwork.* There is a large amount of paperwork beyond technical documentation, drawings, and any documentation generated in the design process itself. Part of this paperwork is actually inherent to the process but the rest of it comes from the need to transmit on paper the information that is not shared in person. So, it can be estimated that in these systems only 10% of the documentation is really needed while 90% is just "accompanying papers" attached to the valuable set of design information. Early introduction of information and communication technologies (ICT) didn't help improve the situation as, in general, these technologies fostered the generation of greater amounts of information increasing the difficulty in handling it (Goldratt 1990). In general it can be said that the early approaches focussed on automating the flow of information in its original state, that is to say, with all the old problems and defects.
- *Local restricted vision.* People involved in design have a local limited version of what are they doing and what for. Each organizational unit in the process has just the knowledge of its piece of work but nobody controls the whole history. Top management and marketing should have a "bird's eye" perspective of the whole process but they are generally flying too high. In general they are more interested in figures and deadlines. Furthermore, this overall approach is not correctly transferred to the operative people.

[4] The acronym DFx stands for any of the techniques supporting the design approach to manufacturing (DFM), assembly (DFA), maintenance (DFMt), disassembly (DFD), and others.

- *Methodologies and tools.* There are many of them, including ICT tools but generally speaking their implementation level is not too high and they are used in an isolated manner, creating disconnected islands of technology.

1.4 Tools to Integrate

In this section, some of the most relevant tools to be used in a design process are going to be presented and their integration in the whole process will be discussed later on.

The following list (Table 1.2) presents a more extended list of tools. Not all of them are presented in this section but some of them will be discussed when introducing the new process.

Table 1.2 List of tools

Tool	Product design	Process design	Management
6σ (Six Sigma)	Design for 6σ (DFSS): this term is used as intending to design the product aiming to achieve a 6σ process, that is to say "zero defects" which in practice means failure rates in parts per million (ppm). In fact DFSS encompasses many of the tools described here	Continuous improvement in the production processes by statistical means intending to achieve "zero defects" (in fact ppm)	Continuous improvement
9 S			Tools for continuous improvement in performance
Benchmarking	Analysis and incorporation of best practices in competitive or similar products	Analysis and incorporation of best practices in similar processes	Analysis and incorporation of best practices in business

Table 1.2 (continued)

Tool	Product design	Process design	Management
Capacity studies		Analysis of the capacity of the processes to produce the required parts to the specification	
Design of Experiments Taguchi techniques	Definition of the value of key parameters in order to achieve a "robust" design	Definition of the value of key parameters in process in order to achieve a sound manufacturing process	
Feasibility studies	Analysis of the possibility of producing the product with the available means	Changes required in the current manufacturing processes	
Failure Mode and Effect Analysis (FMEA)	Analysis of potential failure modes in the product	Analysis of potential failure modes in the process	
Function Analysis System Tree (FAST)	Analysis of the operation mode of the product through its functional structure	Analysis of the operation mode of the process through its functional structure	Analysis of the operation mode of the management processes through its functional structure
Kaizen		Small continuous improvement in process	Small continuous improvement s in the quality assurance system

Table 1.2 (continued)

Tool	Product design	Process design	Management
Kansei Engineering	Integrating customer's perception in the product		
Make or buy	Decision on what parts to make or buy		
Process Decision Program Chart (PDPC)			Flow chart with the sequence of decision–making in case of undesired or unexpected events
Poka-yoke	Design preventing failures (making impossible their occurrence) in production or assembly phases	Process design making the failure impossible in manufacturing process	
Quality Function Deployment (QFD)	Deploying product design starting from customer's requirements and needs	Deploying process design adapted to product requirements (coming from customer's needs)	Could be applicable to any business process but not so evidently
Quality Costs Control	Feedback to design improvement from failures in process	Feedback to continuous process improvement from failures in process	Continuous improvements in the overall process

Table 1.2 (continued)

Tool	Product design	Process design	Management
Reliability, Availability, Maintainability techniques (RAM)	Statistical techniques to improve product performance, life expectations and failure rates	Same to manufacturing processes	
Single minute exchange of die (SMED)		Techniques to design the manufacturing process enabling tools quick change-over	
Statistical process control (SPC)		Statistical process control to assure fulfilling specification	
Theory of Constraints (TOC)		Aiming to elimination of process bottlenecks	Aiming to elimination of bottlenecks in business processes
Theory for innovative problem solving (TRIZ)	System for systematic innovation in product	System for systematic innovation in process	Systematic innovation business system
Value Analysis (VA), Value Management (VM), or Value Engineering (VE)	Costs to value balance in the product components	Costs to value balance in the process phases	Costs to value balance in the business processes

1.4.1 Feasibility Studies

Feasibility is the assessment of the possibility that a design, process, or material for production fulfils all the engineering requirements with the minimum capacity required at the specified volumes. Feasibility assessments are required for new products, changes to products and processes, or important changes in volume. For these evaluations, planning tools such as Failure Mode and Effect Analysis (FMEA), control plans, process capability analysis and design of Experiments processes are used. Manufacturing feasibility should be established before taking on any commitment with regard to tooling or manufacturing resources.

Consequently, there are three concepts that should be taken into account when talking about feasibility:

- Technical requirements (specification)
  - Product
  - Process
- Production volume
- Profitability

These three concepts are very closely related to each other: The product has to be manufactured fulfilling the required specifications with adequate production processes capable of delivering in time the volume needed to be profitable in the market in the medium/long term.

The relationships and responsibilities of the teams working in the development process should be planned and adapted on a case-by-case basis.

One important concept in the feasibility studies is that of *bottlenecks*. A bottleneck is faced when any of the current processes is incapable of accomplishing the objectives established in terms of cost, quality (*process capability*), manufacturing volumes, or any other requirements.

The detection of these bottlenecks in the design phases constitutes one of the most evident advantages of working on *Concurrent Engineering* (joint product/process design) as will be discussed in the following chapters.

The bottlenecks mark working priorities since the limited resources of any organization should focus on solving these problems. The possible solution channels will basically emerge from the utilization of some or all the following techniques (some of them will be further developed later):

- Make or buy analysis
- Functional analysis
- Quality Function Deployment (QFD)
- Innovation
- Investments analysis
- Ongoing improvement

- Computer simulation
- Benchmarking

The fundamental aspects to be taken into account when these analyses are performed are:

- Optimization of the cost/performance ratio
- Clearance of wastage: in the product (the customer only pays for what he/she wants); in process (anything which does not add value to the product)

1.4.2 Make or Buy

Deciding whether a new product is to be introduced into the current portfolio is possibly one the most important decisions to be made.

The analysis should be made for each of the elements of the product family tree. Obviously, the decision is based on the estimate of whether we can make a given component at a cost that renders it possible to obtain profits for the given selling price.

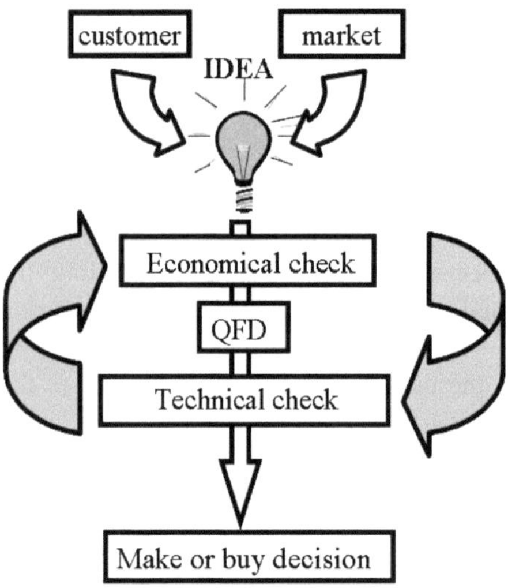

Fig. 1.7 Make or buy analysis

Make or buy analysis is an iterative process as can be seen in Fig. 1.7. Starting from the "idea" coming from individual customers (or clusters of them), the market, or both, economical (pay-off) and technical (feasibility and capability) checks have to be done. QFD methodology (see Sect. 1.4.3) lies in the middle of the iterative circle as a methodology that can push overpassing current limitations beyond the current status by incorporating innovation.

Iterations will finalize in time due to time pressures (not desirable) or to economical or technical clarification (go/not go).

In most cases (not to say all of them) the selling price is predetermined, either by the market which establishes the price bracket in which we may operate conditioned to the target market, current and future competition, socioeconomic environment, *etc.* or else because the customer has commissioned us with developing to a given price.

Modern design trends talk about the "*Design-to-Cost Objectives* " (DCO) methodology that consists of designing, from the outset, on the basis of a fixed maximum acceptable cost for all the phases: *concept, development, industrialization, production and commercialization* plus for the finished product broken down to the level of the parts of the family tree. Evidently, DCO will be set with a level of improvement over the recorded historic data, but it must be a reachable goal based on a rational and serious estimate.

This process therefore entails a detailed cost study including development costs and which, at part or component level, should take into account material and manufacturing costs.

Focusing on the specific case of the "make or buy" analysis, in the DCO an objective cost will be assigned to each part, and therefore the internal manufacturing (make) or purchase (buy) decision will be determined by the estimate of own manufacturing costs *vs.* market costs.

In principle, products, components or parts under analysis could simplistically be classified as:

- *Standard.* Those that are easy to find on the market (they can be bought in the *corner hardware store*) and in many cases are standardized. For this type of parts it is relatively simple to find reliable suppliers and negotiate a suitable price with them.
- *Mature technology.* One step below the preceding ones which, without being standardized, require a type of mature technology that is very widespread on the market and has therefore been mastered by many manufacturers.
- *Specific to our technology.* Parts well fitted to the company development and manufacturing systems (*know-how*). The company has regularly been producing similar parts and therefore it should be relatively easy to make the necessary changes to adjust to the objective cost, while on the other hand it is not easy in the short term to find alternative suppliers.
- *In-between.* In this group, there will logically be included parts situated between the last two categories, which could be sub-contracted to a specialized supplier or be manufactured in-house. This last decision, in any event, would normally require a major investment in technology and production resources, manufacturing specialized personnel, *etc.*

Initially, the efforts to improve simultaneously part design and their production process should focus on the parts of the last two categories.

In this case, the use of other tools such as QFD (Sect. 1.4.3), Value Analysis (Sect. 1.4.6) and creativity and innovation techniques, TRIZ among them (Sect. 1.4.4) are highly useful for developing different alternatives.

Other tools that should help to make decisions are:

- *Group technology.* Redesign of parts which, while they are apparently different, share common elements (diameters, manufacturing parameters, holes, threads, *etc.*) which allows them to be produced on the same machines and on flexible manufacturing lines.
- *Design range.* Design new parts and components envisioning their future addition to new designs following the previous criterion. This process affects the current cost of design but is a future investment and therefore will be taken into account in calculating the DCO for future designs. Evidently, there is always the risk that these *future designs* do not come to fruition, whereby the "investment" would have been useless. Section 3.5 deals about risk in innovating in products, where these issues will be discussed in deep.

The use of these techniques, combined with the inclusion of common parts (even standard) and a modular design, greatly cheapens design (design is common and parametrizable) and manufacturing costs, and moreover in many cases permits the use of the current production media.

Evidently there is always room for extreme alternatives that entail strategic decisions which in some cases really need to be made:

- *Becoming a pure distribution company* buying and sub-contracting everything
- *Introducing a new technology* (or improving current technology) and manufacturing everything at home

1.4.3 Quality Function Deployment

Quality Function Deployment (QFD) is a methodology that started in Japan by the end of the 1960's (Akao 1990), jumped to USA in the 1980's (King 1989; Mazur 1993; Terninko 1996) and expanded from there to many other countries in the world (Sorli and Ruiz 1994).

QFD systematically translates customers' requirements (voice of the customer) into design requirements (voice of the engineer). QFD is a proven approach to improving customer satisfaction, reducing product development cycle times, integrating internal and external suppliers, lowering start-up problems, and developing a customer-driven knowledge base.

QFD may be defined as a "Structured process aiming to gather the voice of the customer, translate and integrate it into life cycle product/process design requirements with contribution of all actors in the development process".

Analyzing this definition, the following main characteristics may be pointed out:

- Strictly speaking QFD is not a tool but a methodology which is used as an aid for structuring and systematizing a series of steps and operations which are commonly carried out in a disconnected and poorly ordered manner.
- Namely, this requires a significant change of mentality in the current managerial culture. Specifically, the correct application of QFD promotes the participation of representatives of different functions from the start of design process, contrary to the traditional way of working in isolated departments and through successive stages.
- In each matrix or stage of QFD project, the dynamics internal to the process "suggests" the use of other tools as an aid to channel, extend, select, *etc.* the information available or discover which is missing. The QFD handles *information*, helps to structure it, classify it, determine priorities and, especially, quickly discover gaps to be filled in.
- The logic process of QFD opens new paths with the aid of other tools, as mentioned above. Some of these paths could be ineffective (low profitability, unsuitability of technology available, *etc.*), and the others more or less valid. After reaching this point, a strategic decision should be taken in order to concentrate efforts in a single direction. However, today discarded paths due, for instance, to current existing limitations or to any other cause, may be recovered in the near future upon changes in the situation (*i.e.*, new technological developments). QFD stores all these information allowing easy updating and revision of past decisions. This is one of the great potentials of QFD.

In brief, QFD "forces" to follow strictly a series of systematized stages, in which all functions will participate. In this way, the following goals are reached:

- Nothing is "missed." The system tries to foresee all product requirements for its entire life.
- All information is channelled toward the customer – *The voice of the customer*– giving real sense to the motto "The customer is the king".
- The creativeness of team members is enhanced. Several techniques are used to try to "invent" new performances of the product to give a greater satisfaction to the customer and to allow the manufacturer to achieve a higher competitiveness level.
- A series of tools like those mentioned in Table 1.2 – scarcely used in general – are applied in a joint and systematized manner.

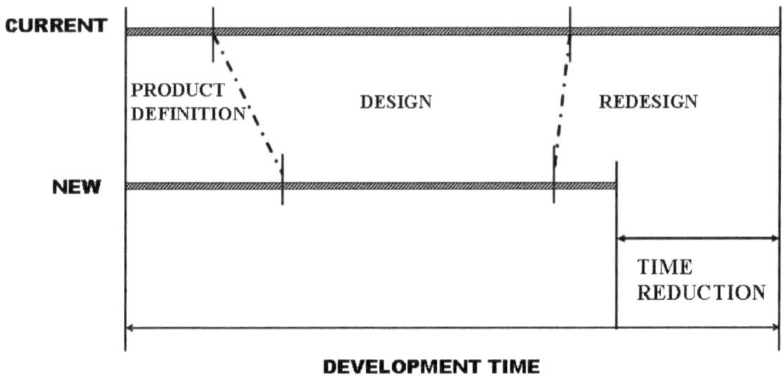

Fig. 1.8 QFD supported design process

Benefits of QFD. The aspects mentioned up to now are clearly benefits resulting from the application of the methodology, but it can be considered that the most significant advantage of QFD is *its high competitive character*: the correct use of QFD methodology will allow the company to reach a privileged position among the competitors. This fact emerges from aspects such as:

- *Differential performances.* As stated above, the QFD promotes and enhances the creativeness of team members to search for brilliant ideas which will add new and valuable characteristics or performances to the product. These novelties will gain an important market share before the competitors take suitable countermeasures.
- *Reduction of development time.* See Fig. 1.8. Development time also known as launching time or "lead time" will be greatly reduced as a result of working within a multidisciplinary team in which several functions participate, having a large amount of information available. The product definition and design stages are shortened as compared with the traditional working procedures and, especially, the redesign stage almost disappears. In contrast, in the current system the product commonly passes through several redesign stages as it proceeds through the different involved departments: procurement, production, quality, *etc.*, all problems appearing inherent to a design exclusively made in the development engineering department.

  In summary, the product will be introduced earlier in the market, increasing the possibilities of success and preventing operation faults.

- *Reduction of engineering changes.* As a result of what has been pointed out up to now plus the characteristics mentioned above of *missing nothing* and *participation of all functions,* design is carried out in such a manner that the product will meet any requirement specified by any *internal or external customer.* As most (hopefully all) of the requirements are integrated in the specifications from the very beginning of the design, engineering changes are significantly reduced and required changes are concentrated in the previous definition and design stages where changes are less costly. This point is much supported by the concept of "Concurrent Engineering" to be discussed in further detail in Chap. 3.

- *Cost reduction.* This is a direct consequence of all aspects and, though of the greatest importance, it is not worth making further comments, as it is obvious.

1.4.4 Theory for Inventive Problem Solving

The work on the Russian Theory for Inventive Problem Solving (TRIZ)[5] (Altshuller 1992) was begun by Genrich Altshuller in 1946, and was continued since then by his students and followers. Many of them immigrated to the USA later on and have continued working on it from then on.

The figure of Altshuller[6] actually arouses a high level of admiration among his collaborators, disciples, and followers. Feeling that can actually be felt in any of the series of the "International Congress on TRIZ and its applications" that are run yearly in the United States, organized by the Althsuller Institute for TRIZ Studies[7] sponsored and attended by many industrial companies worldwide.

TRIZ, as Altshuller developed it, is based on some fundamental premises that can be enunciated as:

- *Systematization.* The "inventor" follows a method but it is intuitive. However TRIZ systematizes it.
- *Solutions.* The solutions are given but must be found which can actually be done.

---

[5] TRIZ is the Russian acronym for "Theory for Inventive Problem Solving".

[6] Genrich Altshuller (1926–1998) was born in Tashkent (old URSS) and died in Boston (USA). He was a successful young inventor and made his first invention at the age of 14. In 1946 he developed his first mature invention, a method for escaping from an immobilized submarine without diving gear. This invention was immediately classified as a military secret and Altshuller was offered employment in the patent department of the Caspian Sea Military Navy.

In 1948 he wrote a dangerous letter: "Personally to Comrade Stalin." The author pointed out to his country's leader that there was chaos and ignorance in the USSR approach to innovation and inventing. "There exists a theory that can help any engineer invent. This theory could produce invaluable results and revolutionize the technical world." Altshuller and his former schoolmate, Rafael Shapiro, were asked to come to Tbilisi (Georgia) to receive a National Competition Award in 1950 where they were arrested instead. Charged with "inventor's" sabotage, they were sentenced – as was usual in those days – to 25 years imprisonment.

Against all logic, Altshuller exploited this experience by working with highly experienced people who were there in the same distressing conditions. In that way, he continued developing and testing his theory which got higher impulse when Altshuller was finally liberated after Stalin's death.

[7] http://www.aitriz.org

- *Patterns of evolution.* Technological systems do not evolve randomly but following some evolution patterns or guidelines.

TRIZ is also based on the following principles:

- *Ideality.* All technological systems move toward ideality. Maximum ideality level is reached when a system performs its functions without actually existing.
- *Contradictions.* Any inventive problem contains a contradiction. If this contradiction can be solved, innovation appears.
- *Systems approach.* A system is not isolated but formed by *sub-systems* and integrated in a *super-system*. In many cases the real problem is not where it seems to be but in any of the other levels. The solution in another level may be easier to find and less costly to implement.

In accordance with the experience accumulated up to now, TRIZ is an excellent tool for innovation. TRIZ has been used successfully for the development of new products, for the forecast of future product developments or evolution of technologies, for building patent “fences,” for uncovering the causes of past failures, as well as identifying and eliminating potential causes for failure prior to their appearance.

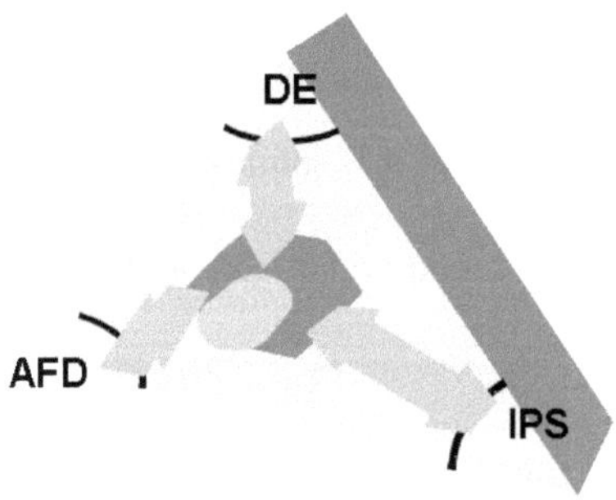

- **AFD: Anticipatory Failure Determination**
- **DE: Directed Evolution**
- **IPS: Inventive Problem Solving**

Fig. 1.9 TRIZ three basic lines of application

Ideation/TRIZ Methodology improves and extends the classic TRIZ (Arciszewski and Zlotin 1998). It has three basic working lines (Fig. 1.9):

- *Innovative Problem Solving (IPS).* Problem solving by generating a high number of conceptual ideas in limited time spans. The process follows the denominated ARIZ[8] algorithm for systematic problem analysis and idea generation.

[8] ARIZ is the Russian acronym for TRIZ based algorithm.

- *Directed Evolution (DE).* Analysis of the evolutionary lines of product, service or organization; forecast and identification of future trends, creating business opportunities, finding new substitutive products, *etc.*
- *Anticipatory Failure Determination (AFD).* Bases on ARIZ algorithm to generate ideas to "invent" failures before they show up, being able to "determine" the failure anticipatorily. Another branch of it supports creative Failure Analysis (FA) once the failure has occurred.

Plus another couple of powerful applications:

- *Value Analysis.* Applying the "Idealization" concept and utilizing TRIZ tools for ideas generation (basically IPS and AFD) Value Analysis (VA) may be enhanced and reinforced (see Sect. 1.4.6)
- *Revealing new (hidden) business opportunities* by the use of DE

Finally, as can be seen in Fig. 1.10, TRIZ gives a high potential at the stage where practically there does not exist any traditional tool: *finding solutions.*

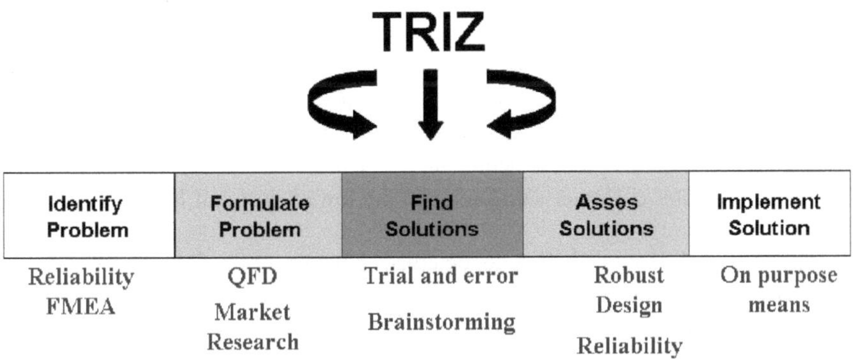

Fig. 1.10 Added value of TRIZ

1.4.5 Failure Mode and Effect Analysis

Failure Mode and Effect Analysis (FMEA) is a method which, by means of a systematic analysis, contributes to the efforts of identification and prevention of failure modes of a product or a process, evaluating their seriousness, occurrence, and detection, to prioritize the causes on which action must be taken for preventing occurrence of these failure modes (Stamatis 2003; Dailey 2004).

In other words, FMEA makes it possible to identify, in a simple way, the significant variables of a process/product that allows us to determine and establish the corrective actions needed to prevent the failure, or the detection thereof, if it occurs, ensuring that faulty or unsuitable products do not reach the customer or – in the case of processes – they are not put into operation.

The people that perform FMEA on a system should be familiar with the failure modes of each part. Moreover, they should investigate the possible causes of each failure mode and determine, for each failure mode:

- *The seriousness.* Or consequences of the effects of the failure
- *The occurrence.* Or frequency of appearance
- *The probability of non-detection.* What are the means for identifying the failure (signaling, alarm, test or periodical inspection) before the effect shows up
- *Corrective actions.* What possible corrective actions should be applied in order to prevent or reduce their appearance

The combination of the figures of *seriousness*, *occurrence*, and *non-detection* produces a so-called "risk priority number" (RPN) which helps prioritize the implementation of the *corrective actions* (See Sect. 2.5.2.3).

The FMEA is therefore a highly useful tool in product development that makes it possible to ensure systematically that all the potentially conceivable faults have been taken into account and analyzed.

FMEA is based on the work of a multidisciplinary team that includes all the departments affected, although specialized experts (from inside or outside organizations) should be included in highly complex cases.

The application of the FMEA, in a word, fosters active participation of all the agents involved in the process achieving an increase in creative potential finally resulting in greater customer satisfaction at the lowest cost and from the first unit produced.

The FMEA is, in a nutshell, a documented summary of the thought process that takes place in the mind of the engineer or designer of the product or process being developed or manufactured. Naturally, it is based on the designers' experience and on the accumulated knowledge on past problems. It is a systematic approach that formalizes the mental discipline that the best designers always apply.

1.4.6 Value Analysis

Value Analysis (VA) is a system of techniques and procedures geared towards identifying superfluous costs. The analysis of value (sometimes referred also as "Value Management" – VM) is not just a conventional cost reduction method, but is rather a more perfect procedure that makes it possible to achieve results with greater scope (Réfabert and Litaudon 1988; Miles 1989; Litaudon and Réfabert 1992; Shillito and De-Marle 1992; Luque and Montoya 1995; European Commission–DGXIII 1995).

Important definitions in *Value Analysis* are:

- *Functions.* The functions of a product may be regarded as the services that may be rendered to the user, including such aspects as the aesthetic satisfaction it provides among others.[9] Each *function* has:
  - *Value.* Contribution of each function to the global satisfaction provided by the product to the user
  - *Cost.* The cost of a function is defined as the minimum absolute cost needed to perform a function

- *Value Analysis.* It analyzes the balance between the function performance quality and the function cost seeking to achieve the highest performance (value) at a cost as close as possible to the value they provide. This aspect is explained next.

Value Analysis sees the product not as a conjoint of elements and materials but as a system that will perform a series of functions needed and appreciated by the user.

"Function Analysis" is performed by using tools such as "Function Analysis System Tree" (FAST) (Akiyama 1991) which allows the analyst to decompose the product operation mode through a tree of interrelated functions aiming to the achievement of the main function expected by the customer.

Estimation of *value* per *function.* The importance that each of the *functions* has for the user is called *value* (the "value" it provides to the user) and is evaluated based on information collected from real consumers. At the end of the analysis, each function has got a relative percentage weight in the global (100%) customer satisfaction.

Calculation of *cost* per *function.* A thorough analysis of the product costs – broken down by components – is performed. The contribution by component to the execution of a *function* is estimated in percentage terms (*e.g.,* function X is performed by component A in 20%, component B in 45% and C in 35%). *Function cost* is then calculated as the summation of individual component costs by its contribution percentage.

Finally the method compares the cost of each *function* (in percentage of the total cost) with the *value* it provides to the customer (also in percentage). In an ideal situation the percentage of cost should be equal to the percentage of *value* to Customer for each *function.* Therefore big imbalances (high costs *vs.* low satisfaction level) are the objectives for cost reduction.

When this technique is applied to manufactured products it is called *Value Analysis (VA)*. New denomination of *Value Engineering (VE)* highlights its potent applicability either to a new product in its conception or design phase, to manufacturing or business processes or to services.

---

[9] The concept of "function" should not be confused with the "solution" provided. For example, the function performed by a screw is to "secure" and not to screw, since the latter is the solution given to the secure function.

1.4.7 Design of Experiments

Design of Experiments (DoE) is intended to find the best combination of factors that will optimize the output (in quality and quantity) of a manufacturing process or the performance of a system (Box *et al.* 1978; Hicks 1982; Montgomery 1991; Hinkelmann and Kempthorne 1994; Barrentine 1999; Placzkowski 2000).

When analyzing the way a process may be optimized, the factors that influence it must be taken into account. The first step is to fix the levels in which the factors are to be studied (high and low values); these factors must be selected based on the experience of the people involved in the study, the most accurate the factors, the better will be the results of the study. The combination of a level of each one of the factors being studied is denominated *an experience*. The *DoE* is the conjoint of these experiences. All experiments have a random component, *noise* (caused by the uncontrollable factors), which is a variability of the response with the factors under control.

In any industrial processes a distinction should be made between:

- *Variables.* Those variables affecting the process that may be controllable (*i.e.*, temperature, pressure, cooling oil flow, reaction time, concentration of additives, *etc.*) or uncontrollable (outside temperature, weather conditions, level of humidity, employees behavior, differences in material quality, *etc.*).
- *Results.* Those whose value comes from the conjunction of all the process characteristics *i.e.*, weight, fat content, process yield, degree of whiteness of the matter, *etc.*

Those belonging to the first group are called *factors* that affect the process. Those belonging to the second group are called *responses*.

- *Levels* are the different possible statuses of a *controllable factor*. The most used levels in the Design of Experiments are a low level (–1) and a high level (+1), for a two-factor level; a low level (–1), a medium level (0) and a high level (+1) if we are studying a three-level factor.

Traditional methods operate by moving only one factor each time, maintaining the rest constant, *i.e.*, changing only the level of one factor and leaving the others on a fixed level. This system has a great level of uncertainty since it does not allow one to know what may happen if the experiment is run moving the other factors or changing the level. This uncertainty could only being reduced by running a huge number of experiments.

Design of Experiments on its way provides a means to solve both the typical errors of the experimentation and the limitations from traditional methods. It allows movement of all the factors *simultaneously* in a factorial design: a design where all the factors intervene in all their levels that affect response. As a result:

- It is possible to study the factors with few experiments

- We get more information (interactions)

The objective of the *DoE* is to delimit the field of experimentation selecting the levels on which the factors should be fixed to be able to study their influence on response.

The variation caused by the change of level of the factors on the response is the *effect*. A good *DoE* should make it possible, with the lowest possible number of experiments, to obtain the maximum amount of information on the effects.

In such a way we obtain an empirical model that relates the factors to response. This model will allow deducing the optimal working conditions *(values of the factors)* that will yield the best technical and economic result minimizing the influence of the uncontrollable factors. The achievement of a *robust design* – that is to say a design (of product or process) that can be considered mostly insensible to the *noise* or uncontrollable *factors* – is then enabled.

Design of Experiments encompasses a series of techniques such as the following:

- Complete factorial designs
- Fractionated factorial designs
- Latin square designs
- Analysis of variance (ANOVA)
- Regression analysis
- Evolutionary designs

As an example, following a *DoE* it is possible to study three factors (X_1, X_2 & X_3) with eight experiments (responses Y_1 to Y_8). By means of combinatory methods, eight possible combinations with three factors at two levels each are achieved, giving the data in Table 1.3.

Table 1.3 Experimentation table

X_1	X_2	X_3	Response
–	–	–	Y_1
+	–	–	Y_2
–	+	–	Y_3
+	+	–	Y_4
–	–	+	Y_5
+	–	+	Y_6
–	+	+	Y_7
+	+	+	Y_8

The main benefits of the Design of Experiments may be summarized as:

- It saves time and money
- It makes it possible to experiment with fewer process interruptions

- It statistically discards the effects of undesired variables
- It indicates, scientifically, how many data should be collected and what variables to include

The main risks or hazards are:

- Unless planning is good and the results are studied statistically, the conclusions may be wrong or give rise to false clues
- In non-statistical experiments, many of the effects observed tend to be mysterious or inexplicable
- Time and effort may be wasted by studying the wrong variables or obtaining too much data

1.4.8 Taguchi Techniques

Taguchi techniques may be regarded as a variant, more engineering and less mathematic, of the DoE (Byrne and Taguchi 1987; Taguchi 1987; Box 1988; Taguchi and Chowdhury 1999; Roy 2001).

Taguchi's main contribution is the development of tables with a reduced combination of experiments claiming that the accuracy of the results is significantly similar to the one obtained by a whole DoE as explained above.

Another interesting approach from Taguchi's works is the use of the concept of the signal/noise (S/N) ratio from electronics adapted to the factors in the experiments. Noise factors are those difficult to control so increasing the ratio S/N means that important factors influencing the process (signal) are "robust" enough to the noise factors (Box 1988). The concept of "robustness" in Taguchi's approach means setting the value of the factors in such a level that will make them insensible to the uncontrollable variations (noise).

In that way, his technique is considered as being more adequate to industrial environments where the easy and rapidity of obtaining good results in short time and less cost is more important than the achievements of sound scientific experimental results characteristic of academic and research environments.

1.4.9 Process Decision Program Chart

Process Decision Program Chart (PDPC) (Brassard 1989) is a flow chart (Fig. 1.11) collecting all possible undesired situations along the implementation of a plan with the aim of:

- Establishing preventive actions impeding their occurrence

- Foreseeing corrective actions or alternative ways of evolution to be used in the case of the undesired event finally appearing

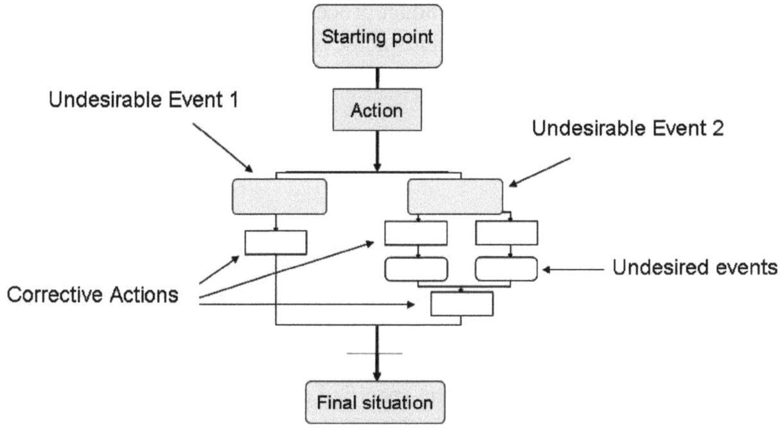

Fig. 1.11 PDPC flow chart

It is recommended to use PDPC in cases such as:

- *New processes* with unknown results are being set up
- *The plan is complex*, critical and it has high failure probabilities
- *Anticipation to difficulties* is desirable

PDPC should be used in the design process related to Risk Management (see Sect. 2.5.2). The process for design and development of new products fulfils most of the characteristics mentioned above:

- In many cases, it is a new uncertain process
- Complex, critical process. High failure probabilities that should be reduced with a consistent and coherent process supported by efficient tools
- Desirable anticipation to difficulties

1.5 Tools for Continuous Improvement

The next set of tools is intended to be used in order to continuously improve already existing either products, processes or systems. In consequence they work on something physical and they are hardly applicable to virtual inexistent systems as is the case in new product and process development. However in many cases, "new products" (as will be discussed later) can be based on incremental innovation carrying on improvements in existing products. In that case, these kind of tools are useful and a "must" in some cases.

On the other hand, knowledge feedback from continuous improvement has to be managed, stored and used in new product/process development in order to introduce into them all relevant improvements achieved within time.

1.5.1 Capability Studies

Whenever a manufacturing process is being designed, set up, or has suffered important modifications, a *Capability Study* is needed in order to foresee the ability of the process to manufacture the parts it has been designed for at the required level of quality.

So, Capability Studies are quite relevant in order to assure that the process will be capable of manufacture the parts to the specified quality at the required production output.

The quality level is assessed by taking samples of the parts during regular production runs and measuring in them the previously identified critical parameters (see Sect. 1.5.2 on Statistical Process Control).

Process capability. First of all, the process has to be brought *under control,* meaning that the process has already achieved a regular operation status and all set up fittings, problems, and adjustments can be considered as finished.

The process is under control when only common causes of variation appear: minimal differences on raw material, working shifts, *etc.*, normal variability due to the ineluctable fact that it is impossible to obtain two exactly equal parts to the minutest detail.

Definition. Once the process is under control *Process Capability* may be defined as "the amplitude of the interval of variability of individual observations."

Therefore, a process capability is the relationship between the range of the specifications admitted as acceptable and the dispersion of the process where the lower its dispersion the greater its capacity will be. In summary, process capability represents the long range variability of the process.

The capacity of a production process can be broken down into:

$$\text{Process capability} = \text{Short range variability} + \text{Other sources of variation} \tag{1.1}$$

where the short range variability is based on the machine capability whereas other sources of variation comes from changes in operators, different shifts, variation in material, *etc.*

Process capability index. The process capability index (C_p), is defined as a measure that relates and compares the specifications and tolerance of a process, so large C_p values indicate a more comfortable situation that can be achieved with:

1. Broad limits of specification
2. Small dispersion

In a quality improvement program the analysis of process capacity is a vital part; its main benefits include the following:

- To predict how the process will remain within specifications
- To help the engineers/designers to select or modify a process
- To help to establish a sampling interval for process control
- To specify performance requirements for new equipment
- To make a proper selection between different competent suppliers
- To plan the production process sequence when there is an interactive effect of the processes in the tolerances
- To reduce variability in a manufacturing process

For this reason, the analysis of process capacity is a technique which is widely applied in many segments of product life cycle; for example, it can be applied to product design, supplier agreements, planning, production, *etc.*

1.5.2 Statistical Process Control

Statistical Process Control (SPC) means using statistical techniques to control the production process with a view to limit the variability of quality characteristics and avoid deviations in them. It is based on forecast by means of the probability function. This technique includes tools such as control graphs, sampling and pre-control plans (Shewhart 1939; Deming 1975; Wheeler and Chambers 1986; Oakland 2002).

SPC should be used in a continuous base (*routine study*) on any process in order to assure the required quality level in the products produced by this specific process. On the other hand, whenever a change is introduced in a process as in the case of incremental improvements in product/process (*new product/process*), an *initial study* has to be accomplished in order to make sure that the changes are "accepted" by the system. That is to say that the modifications in the process will achieve the desired impact: improvement.

Differences in the two types of studies can be seen next:

- *Initial study*
  - Normality study
  - Stability study (control graph)
  - Capacity study

- *Routine study*
  - Stability study (graphs)
  - Capacity study

There always exist variations in any working processes, and the fewer the variations (*more stability*), the better the quality of the product. Therefore the clients will be more satisfied with our products or services. There are two types of causes for process variation:

- *Assignable causes.* Also known as *special causes*, they are sporadic factors that destabilize the system. Their identification is immediate and easy, requiring immediate action.
- *Common causes.* Also called *natural causes*, they are factors that may slightly affect the variability of the system. Their presence is random, and they are not easy to detect. If the system suffers continuous destabilization they would require an in-depth analysis.

A *control graph* is a diagram used to check whether a process is in a stable condition or to ensure that it remains stable. In statistics, a process is said to be stable (or under control) when the only causes of variation present are random causes. Although there are different types of control graphs, they all present a similar structure to the one shown in Fig. 1.12.

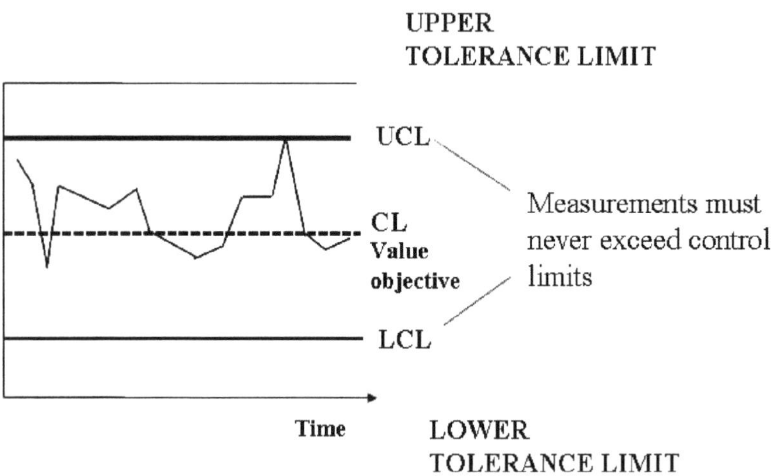

Fig. 1.12 Control graph

On the basis of the information obtained by sampling at given time intervals, the control graph defines a "confidence interval" (see below); if a process is statistically stable, 99.73% of the time the result will be within this confidence interval.

The structure of control graphs contains a "*central line*" (CL) or value objective, a higher line that marks the "*Upper Control Limit*" (UCL), and a lower line that marks the "*Lower Control Limit*" (LCL). The points contain information on the readings or observations made; they may be averages of sets of observations, or their ranges. The control limits mark the "confidence interval" in which the points are expected to fall. Control limits are more stringent that tolerance limits defined in specifications, thus giving higher confidence that any deviation will be detected before reaching tolerance limits (defective part).

We can thus state that a control graph is the graphic and chronological comparison (every hour, day, *etc.*) of the quality characteristics of a product with limits that reflect the capacity, based on past experience, to manufacture this product.

A control graph offers the following advantages:

- It can be used to ascertain the control status of a process
- It reflects production fluctuations, in comparison with statistically established control limits
- It diagnoses the behavior of a process in time
- It indicates if a process has improved or worsened
- It permits the identification of the two sources of variation of a process: assignable causes or random causes
- It can be used as a troubleshooting tool

Two types of control graphs may be distinguished:

- *Control graphs by variables* for measurable data. The most used are:
  - Graph $\overline{X}-R$. Averages and ranges
  - Graph $\overline{X}-S$. Averages and standard deviation

- *Control graphs by attributes* for accounting data. The most used are:
  - *100 p graphs.* Control of the faulty percentage
  - *np graphs.* Control of faulty units
  - *s graphs.* Control of the number of defects per sample
  - *u graphs.* Control of the number of defects per unit

The phases covered in the implementation of a control graph are:

1. Selecting one or several quality characteristics that are to be controlled
2. Noting the data taken from the successive product samples as production is carried out
3. Determining the control limits on the basis of the data
4. Drawing the limits in the appropriate graph
5. Starting to draw the representative points of the production samples after those used for the determination of the control limits on the graph
6. Taking suitable corrective actions when the representative points of the samples are outside the control limits

In all cases, regardless of the type of graph used, the points represented on them correspond to the *mean* values of the samples taken and not to the individual values thereof. The reason for using the mean values instead of the individual ones is simple: mean values are much more sensitive to process variations than individual values.

1.5.3 Quality Costs Control

One of the most precise and reliable measurements of the efficiency of the quality system of a company is the *Quality Costs Control* (QCC). Numerous authors, while disagreeing on how to measure these quality costs, agree that the quality costs in a company account for a very high percentage of the company profits, with figures in the order of 20% of turnover for those companies whose quality management system is not at a proper level (Juran 1962; AFNOR-AFCIQ 1981; Feigenbaum 1991). QCC allows the company to identify critical points where the processes are failing (wasting money) in order to launch adequate corrective actions immediately. In the case of minor innovations in new products/process development, changes in the processes have to be monitored continuously but more in detail in the early phases upon introduction. So QCC is a very valuable tool for so doing. The quality costs incurred by a company may be split into two large categories: cost of failures and cost of the quality system itself.

- *Costs of internal failures.* These are the costs resulting from deviations found before the product is delivered to the customer, ascribable to losses caused as a result of non-conformities detected in the actual company. These costs may include:
  - Non-conformities
  - Reworking
  - Back orders
  - Double cross-checking
  - Obsolete stocks
  - Low machinery efficiency: downtime, maintenance, repair, *etc.*

- *Costs of external failures.* These are costs caused by deviations found after the product has been delivered to the customer; these costs can be ascribed to losses caused as a result of non-conformities detected by the actual customer.

  These costs are the most serious ones due to their repercussions, although at the same they are the most difficult to measure (unsatisfied customers, bad brand image, loss of market, *etc.*).

These costs may include:

- Return of shipments
- Wrong delivery (quantity, dates, wrong labeling, *etc.*)
- Travel to the customer's facility for problem solving (*fire fighting*)
- Other

Besides recording and monitoring failure costs, *failures* should first be settled as soon as possible to prevent propagation; second, improvement should be implemented in order to prevent repetition and finally but most important, failures are a very valuable source of information that should be stored in a repository in order to incorporate the adopted solutions (or better ones) in the design and development of new products.

On the other side the costs of the quality system can be split into:

- *Prevention costs.* These are the costs derived from the efforts to prevent deviations; they are costs ascribable to all activities geared towards preventing the appearance of non-conformities. These costs may include:
  - Training costs
  - Supplier homologation
  - Calibrations
  - Quality planning
  - Preventive maintenance
- *Evaluation costs.* These are the costs derived from an effort to check the quality of the product and the detection of deviations, and are ascribable to the verification of the compliance of products with quality demands; *i.e.,* the costs derived from "searching" for non acceptance. These costs may include:
  - Reception checking and inspection
  - Final checking
  - Manufacturing control
  - Homologation tests and official approvals

It must be stated that, regardless of the category in which a cost is situated (in prevention, evaluation, *etc.*), the most important thing is that this cost should be measured efficiently in an easy way. It should also always be kept in the same cost category so that it can be compared at different times and be used as a quality indicator looking for the set up of corrective actions in order to reduce it.

When establishing a quality cost system, the following tasks are required:

- Selection, identification and description of the quality cost concepts that will be studied
- Method or system to measure the quality cost concepts chosen
- Calculation of these quality costs by using a common base along one financial year

- Analysis of the quality costs by the company Quality Committee
- Approach to quality costs objectives for the next economic year and corrective actions to reach these objectives

By the adoption of corrective measures, the total quality costs will reduce in percentage to the turnover, not only by the reduction itself, but also because less quality failures will cause higher production and sales rates at lower costs.

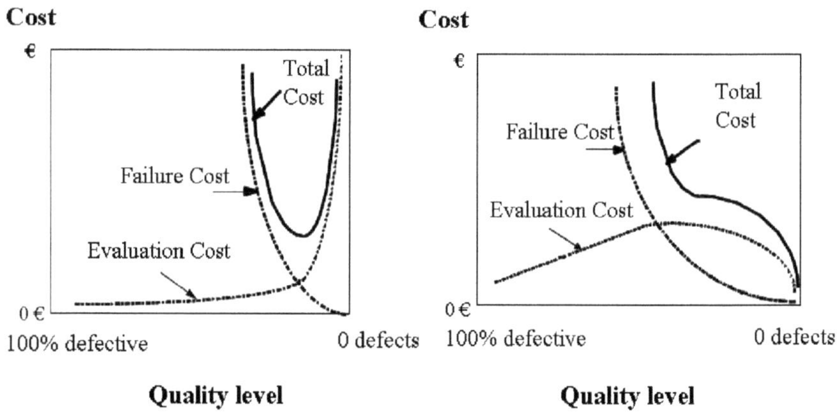

Fig. 1.13 Quality costs

The traditional approach to Quality (left graph on Fig. 1.13) looks for the optimum value for total costs (minimal value) assuming an acceptable percentage of defectives.

The total quality management (TQM) approach, in contrast, aims for "zero defects" (Halpin 1970) which, following the curves on the left-side graph, would increase to infinity the total costs from the sum of theoretical zero failure cost with evaluation cost tending to infinity.

Nevertheless, the real point is that quality costs under the heading of TQM should come to zero converging with a "zero defects" quality level as is represented on the right-side graph in Fig. 1.13.

The rationale for this approach comes from the basic concepts of TQM:

- Quality is the responsibility of each and every member of the organization: self-control
- Creation of a culture of continuous improvement
- Quality Control Department shifts from *control* to *management* and becomes almost a Management Staff Group caring about system updating and catalyzing the efforts for continuous improvement coming from different areas

This means that *failure* costs should become zero by absence of failure and *evaluation* costs become part of the overall overhead cost of the organization.

1.5.4 *Kaizen*

The essence of Kaizen is simple: Kaizen is an ongoing improvement process that involves all personnel, including executives and operators (Imai 1986; Ortiz 2006; Henderson 2006). Kaizen philosophy means "small improvements through continuous efforts."

It begins with the recognition that all companies have problems and have to solve them, establishing a corporate culture in which anyone can freely admit problems exist.

Another important point is that this process is based on team work establishing multi-disciplinary and multi-level teams working together on identifying the causes and delivering solutions. These teams are formed from people close to the workshop on different hierarchical levels and one very distinctive point of Kaizen is to hear and take into consideration ideas and opinions from workshop operators who usually are the most knowledgeable on the problem and the less listened to.

Furthermore, the feeling of being a group raises the commitment and willingness to work on the problem and contribute to its solution.

Kaizen is a customer-oriented improvement strategy: the management should seek customer satisfaction and serve the customer in order to allow the company to survive in the business and be profitable.

Another important aspect is the emphasis on processes. Kaizen has generated a way of thinking oriented towards processes and a management system that supports and recognizes the efforts of personnel to improve processes. The origins of Kaizen in Japan, contrast with western management practices which revise employee performance on the basis of results and seldom reward effort.

Kaizen is related to the aspect of Risk Management (see Sect. 2.5.2) on the side of problem solving and it is also recommended to be used for a continuous improvement process both for company products and processes and for new products design and development.

Chapter 2
Innovation in Product/Process Development

Abstract Innovation is currently understood as one of the most critical factors for success in manufacturing firms. How to achieve real innovation in very demanding industrial environments is actually a very tough challenge. In this chapter, the concept of innovation is going to be discussed, analyzing the main implications of human beings since innovation is clearly coming out of human brains when triggered with some specific motivations or challenges that are not yet well understood from a psychological perspective. Another key issue is the concept of Extended Enterprise (EE) and how to manage innovation within this frame; the change of working paradigm and the new tools needed to enable people from different companies, sometimes distant locations and definitely different cultures to work together, is a new quite important field of research. How to achieve real innovation in new product/process development is also discussed in this chapter. Finally, a section is devoted to the analysis of the risks in innovation in product/process and how to deal with them.

2.1 Being Innovative

In current global markets, innovation is generally one of the most critical factors for success in industrial firms. Former advantages based on aspects such as costs reduction, local natural resources, geographical situation, and so on are not so relevant today since globalization is flattening these issues, and furthermore, needed natural resources are usually coming from outside, thus obliterating benefits of localization. It is a real must to be always aware of the need to foster innovation, fighting against the usual themes such as: "cut your costs", "get focused." Nowadays motto should be "innovate or lose." This new situation needs the introduction of relevant changes in the way the companies work. One of these changes has to be accomplished in the field of new product/process development that is the basis of the success of industrial companies.

Focusing on that, it is very important to know exactly what the discussion is about and a good reference is the Green Book on Innovation of the European Commission (1995) that has elaborated the definition of innovation given in Table 2.1.

Table 2.1 Definition of innovation

To produce, to assimilate and to exploit successfully a novelty in the economic and social spheres in a way that provides inexistent solutions to the problems and allows fulfilling necessities of the people and the society.

This definition, apart from the idea of "introducing something new," brings it the very important concepts of:

- To exploit it successfully
- In the economic and social spheres; that is to say: the market
- Fulfilling necessities of the people and society

The three points could be reworded and summarized to: *to fulfill the necessities of the market* meaning that the real success of any commercial activity will only arise from a good fitting to the market. In summary, innovation may be defined as: "The transition from a novel idea to a successful product in the market" (see Fig. 2.1).

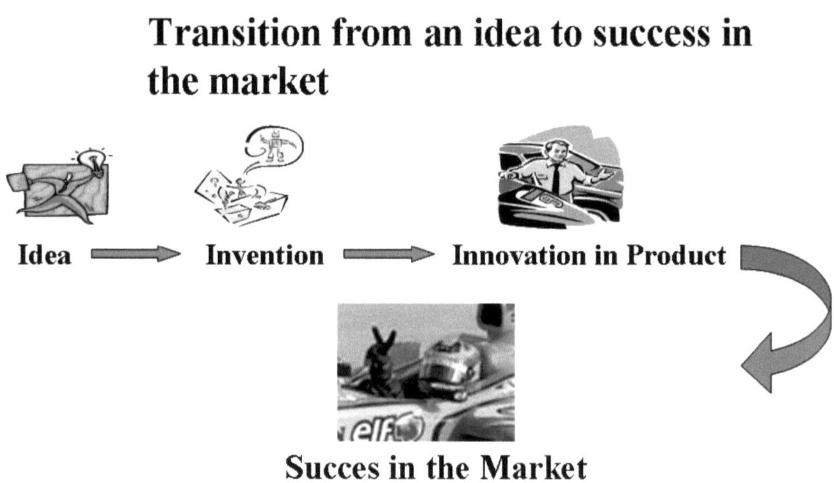

Fig. 2.1 Innovation

Two variables can be distinguished in relation to innovation: *range* and *type*:

- *Range.* Actually there are only two possibilities that may be applied to different aspects of the organization:
  - *Innovation on business management.* It implies working on business processes and covers all areas including strategic management, human resources, marketing, *etc.*
  - *Innovation on product/process.* On the other hand, product and manufacturing (or delivery) processes are currently so interwoven that there no longer seems to be a need to treat them differently.
- *Type.* Two types of innovation:
  - *Radical or breakthrough innovation*
  - *Incremental innovation*

Breakthrough or disruptive innovations should have a significant impact on the business through their impact on the market either by creating a new category of products fulfilling a previously nonexistent demand (Walkman, mobile phones, *etc.*) or by increasing performance level of existing products (injection engines, plasma video screens, *etc.*).

Incremental innovation in its way is very close to the quality field of continuous improvement. Any change in the right direction, adding value to the customer, can actually be considered as an innovation.

Small minor changes in the company's internal processes are difficult to understand as becoming "a successful product in the market" but they are actually also to be considered as "inventions" since, provided they do not have any negative impact through the product to the final customer (better if they have it positive), at least they result in adding value to society in general through such aspects as reducing production costs, improving working conditions, *etc.*

Invention and idea. The origin of innovations is clear: "the great idea (wow)." Even though it may appear obvious, the first main step is to know what are you generating ideas for.

In the industrial world, focusing on product, the right sequence comes from Quality Function Deployment (QFD – see Sect. 1.4.3) (Sorli and Ruiz 1994) as shown in Fig. 2.2.

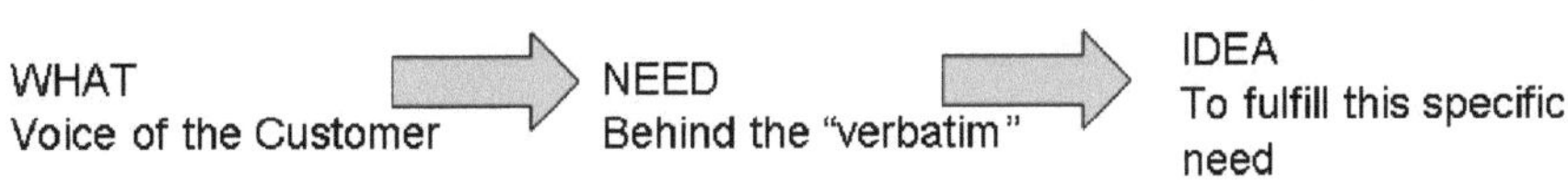

Fig. 2.2 From customer's needs to idea

It may also happen the other way round: the spark for innovation starts with internal dissatisfaction (sales drop, business opportunity, internal/external problems, *etc.*) which through a change plus an improvement becomes an external satisfaction to the customer. From this approach two clear conclusions can be extracted:

- The front end is always the same: *the customer.*
- Innovation means "*change to improve.*" If the change does not bring any improvement it will usually not be harmless; most likely it will carry on disturbances somewhere in the system or, in the best case, will be "good for nothing."

The process for *what's* collection and translation to *needs* is well handled and resolved by QFD. The real gap and challenge is *how to arrive at the breakthrough idea.*

Where do ideas come from? Ideas actually arise only from human brains (Osborn 1942, 1949; De Bono 1967; Altshuller 1992). Good sources (seeds) for them can be found:

- In nature *but* we are unable to notice them
- In tools, artifacts, and devices we use in our daily life *but* are "invisible" for us
- In normal things we are used to do *but* we do not care about them
- In children *but* we do not listen to them
- In other universes *but* we think there is only one: ours
- On the dark side of the moon *but* we always travel to the same side of the moon
- At the end of a long series of "why's" as children use to ask *but* we have lost curiosity a long time ago
- Behind stupid, Utopian, or unrealistic suggestions *but* we dismiss stupid ideas with a frown. Quoting Albert Einstein: "If at first the idea is not absurd, then there is no hope for it"
- Everywhere……but we do not recognize them

Invention. Usually a link is needed between an idea and its practical tangible application to a product or service: the invention.

Without a clear practical objective represented by a "need" from the market place, the application of the "idea" may result in an *invention good for nothing.* This is frequently the case of a technology driven innovation; someone gets attracted by a new technology and immediately looks for where to apply it without really thinking on the key question: "What for?" On the contrary, if the answer is clear, the "invention" will immediately fit into the product/service, achieving the innovation.

2.2 Human Aspects

In the information and communication technologies (ICT) market there exist tools for supporting innovation (*e.g.,* tools supporting collaborative working or idea generation, *etc.* – see Chaps. 4 and 5). New interesting ones are continually emerging; it is for sure that ICT tools will continue growing and will ever increase capabilities and performance.

However, innovation is a serious job that can't rely only on software tools as sophisticated as they could be; there is a real need for methodologies helping people to innovate. Furthermore, innovation means team working which means sharing information. People are in general very reluctant to share information unless they obtain something in return.

Creativity and innovation do not arise directly from the tool itself. As has previously been said, creativity stems only from the human brain and becomes an innovation when applied to solve specific technical problems that will increase the added value to the final consumer. Never forget that only a combination of the three factors (new, successful, adding value) is the real way to achieve innovation.

One of the key resources for creativity is "spare time" to think creatively. Notwithstanding, in current industrial arenas most of the time people are devoting their efforts to perform low value added tasks, fire-fighting, coping with small repetitive problems and nuisances, and in many cases working only for the shake of the organization itself in a much endogamous way. Furthermore, if you try to be creative, the organization may tend to believe that you are wasting your time, a time the company is paying for.

Within this framework, the increasing introduction of ICT tools and the expanding Web facilities combined with the increasing automation of most processes (productive or not) are facilitating the transfer of people's activities from hard manual tasks to soft ones more dependent on intellectual abilities.

As a final consequence, people exercise their mental skills and get more free time, becoming more and more liberated from manual repetitive tasks. The next step is to use this time to be really creative (personal assumption) and to be rewarded by it (company commitment).

2.2.1 Barriers to Innovation

Innovation is not easy to deal with; it has a high level of risk, has to face many difficulties, and needs to be bred and nurtured within a special environment including cultural aspects, means, and systematic approaches.

Some of the generic barriers that have to be overcome for innovating (Piatier 1984) are going to be discussed in this section, while the general approach to "risk in innovation" will be analyzed in Sect. 2.5.

Barriers to Idea Generation. The idea generation process should be divided into two quite distinctive phases as it is graphically shown in Fig. 2.3.

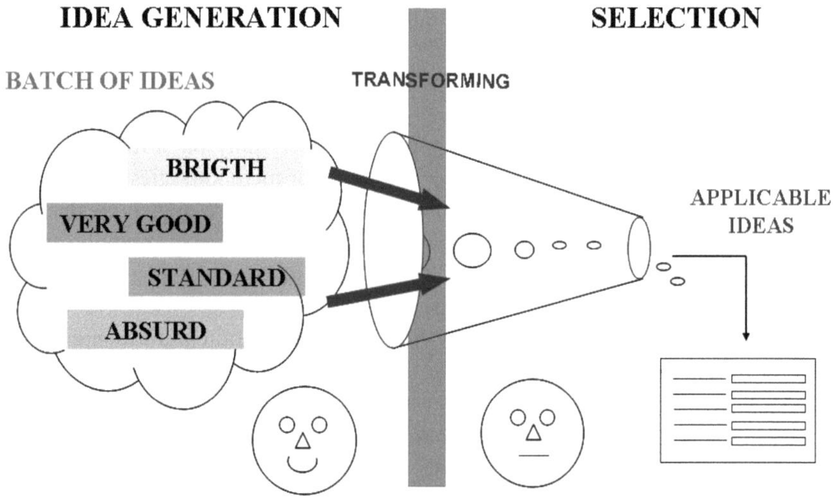

Fig. 2.3 Process for idea generation and selection

The first phase on the left side of the bar is a typical exercise for creativity aiming to free idea generation that could be managed by using a wide range of existing tools (lateral thinking, thinking hats, brainstorming, think tanks, 6-3-5, nominal group techniques, TRIZ, *etc.*) (Osborn 1979; De Bono 1985; Altshuller 1988). It has to be conducted in a freewheeling way, previously creating an open atmosphere, trying to let people evade the day to day routine, and transporting them to a new environment not only in relation to the subject matter but even physically[1], in summary to jump "out-of-the-box." On this approach, people are expected to behave openly, launch a lot of "crazy" ideas,[2] combine, and build upon previous ones, *etc.*

In contrast, the second phase (on the right side) has to be a serious well controlled selection and evaluation process to filter ideas that could finally be applicable. This second part will be discussed in further detail in Chaps. 4 and 5, in relation to the ICT tools supporting this process.

First phase of the innovation process (left side of the figure) has several problems related first, to the difficulties of conducting the process — a good facilitator is a must, and second to the psychological barriers humans build internally to

[1] It is recommendable to use a special room isolated from working areas where some elements fostering idea creation are introduced: colorful painting on the walls, *ad-hoc* furniture, special lights, soft music, *etc.*

[2] As mentioned before: "If at first, the idea is not absurd, then there is no hope for it." Albert Einstein.

block their creativity. These barriers may be of several types (Michalko 1991; Sternberg 1999):

- Perceptual
- Emotional
- Cultural
- Environmental
- Intellectual
- Others

and may appear in different ways such as:

- Self-limitation (perceptual)
- Using stereotypes (perceptual)
- Fear of appearing ridiculous (emotional)
- Not discussing rules (cultural)
- Changing is dangerous (cultural)
- Superficial analysis (intellectual)
- Unique approach (cultural)
- Distractions, monotony (environmental)

Barriers to knowledge sharing. The second phase is a quite longer process (to be discussed in Chap. 5) which needs knowledge handling in order to enable idea analysis, evaluation, combination, and selection of the most promising ones for further analysis and elaboration.

Team work and the use of collaborative tools require knowledge sharing which is one of the main barriers to the process. People are quite reluctant to share knowledge since they feel that it is the main base wherein their values, capabilities and professional status lie. To many persons, sharing and transmitting knowledge mean empowering co-workers (potential competitors) which may eventually imperil their own professional status.

Solutions to foster and improve open knowledge sharing vary greatly depending on the specific circumstances, type of organizations, culture, and other variables. So there are no recipes but a common requirement, a need for an overall cultural change throughout the organization in aspects such as:

- Creating a *win-win* culture: sharing knowledge generates value for all
- Reward systems should be adapted to facilitate *team working* instead of focusing just on individual performance
- *Rewarding innovation* and initiative and *accepting failure.* People should be fostered to try new ways and the organization has to accept they will commit errors
- Empower teams to make their *own decisions* and endorse them from management

Barriers from inside the organization. Organizations need to create a special breeding ground, first to become innovative and second to continue being innovative. Starting a long-distance race – achieving an innovation – though difficult, is easier than keeping on running – achieving an increasing number of innovations in time – which is the brand mark of excellent organizations.

Some barriers hindering innovation will always exist in the organizations. The more innovative an organization, the lower these barriers will be. However it is important to be aware that the barriers can be demolished but if the debris is not cleaned up someone may reuse it to erect the barriers again. If the company's innovation system fails or is neglected, the barriers will grow out of control.

For instance, IBM[3] identifies five barriers to innovation (Andrews 2006):

- *Inadequate funding.* Related to the facts that 1) funding is never enough for innovation; 2) budget time frame does not keep the same pace that potential innovations from ideas arising at any time out of budget
- *Risk avoidance.* People in general do not like changes, are conservative, and are not prone to assume risks
- *"Siloing".* Companies tend to create their own "box" to be enclosed inside and feel protected against the outer environment.
- *Time commitment.* Time (usually long) is really needed for innovation and management hardly accepts that "time for innovating" is paying off. Time is probably the only factor that can never be recovered: "Time is a scarce and valuable commodity."
- *Incorrect measures.* Usual indicators and measurements are not such valid for innovation: payoff is usually longer and there are many intangibles involved.

In fact, some other barriers can be identified as:

- *Organizations not conducive to innovation.* Different types of organizational culture (Kotter and Heskett 1992) greatly influence the disposition of the companies to be or not to be innovative. This in turn, has an impact on the way the organization behaves in relation to innovation on its different levels (management, groups, and individuals), the way the resources are allocated for innovation, how the rewarding systems foster or hinder innovation, *etc.*
- *Leadership.* Innovation starts at the top management level. Since organizations are steered by "persons," the way management behaves and pull the rest of the organization greatly condition the overall company performance (Collins 2001).
- *Employment policy.* Profiles of the people recruited by the organizations will clearly condition the culture of the organization and the working atmosphere which are key issue in the innovation capacity of the companies. For instance Google,[4] renowned for being one of the most innovative companies in our time,

---

[3] http://www.ibm.com

[4] http://www.google.com/intl/en/about.html

looks for a special kind of people to join in which they call "Googlers" – We're always on the look-out for new Googlers – as can be seen in the recruitment Web site of the company.[5]

- *Level of awareness.* Four rising levels of innovation awareness in companies can be considered (Christensen *et al.* 2008). Knowing their own level of awareness allows the companies to shift from one level to another until reaching the highest one:
  - *Unconsciously not innovating.* Not even thinking of needing to be innovative at all
  - *Consciously not innovating.* They know of the need to innovate but….
  - *Consciously innovating.* Slowly innovating without any system
  - *Unconsciously innovating.* Having a system for innovation so innovating without being conscious

Barriers outside the organization. Some barriers can also be encountered outside the organization:

- *Environment not conductive to innovation.* Some environments may hinder innovation though some others may be more prone to innovate more depending on aspects such as:
  - *Characteristics of the company.* For instance, an emerging new technology-based firm (NTBF) (Leonard 2001; Oakey 1994), due to its own characteristics, *must be* more innovative than traditional companies at least in the first years of life
  - *Size.* SME have in general more difficulties in innovating due to scarce resources and time constraints (Pihkala *et al.* 2002)
  - *Sector.* ICT companies, defense, new business (*i.e.,* windmills), *etc.* tend to be more innovative than other traditional sectors.
  - *Regions.* Geographical areas in which the companies are based condition their behavior related to innovation. Silicon Valley in USA is a good referent on how it attracted entrepreneurs from all over the world and launched the "big bang" of ICT companies. In Europe, the Stockholm region in Sweden, Oberbayern in Germany, *etc.* have been classified as advanced regions according to the "Regional Summary Innovation Index" (RSII) (Hollanders 2007).

    These kinds of clustering of companies would generate "The Medici Effect" (Johansson 2004) gathering people with different profiles, expertise, background, training, *etc.* which in general generate a high potential for innovation by taking advantage and great benefits from the synergies out of the "Intersection of Ideas, Concepts and Cultures".

---

[5] http://www.google.com/intl/en/jobs/

- *Administrative and legislative regulations.* Regulations, laws, tax systems may have a relevant impact on the level of innovation of the resident companies.
- *Competition.* Pressure from competition or weak competition has also an influence on the behavior of companies in relation to innovation.
- *Knowledge.* The lower the knowledge on a specific technology, the higher level of innovation is needed to cope with it.

2.3 Extended Enterprise

New ways of working move strongly towards Extended Enterprise (EE). The Extended Enterprise concept in parallel with concurrent enterprising looks for how to add value to the product by incorporating knowledge and expertise from all participants in the product value chain. The semantic difference between both terms will be discussed in the following paragraphs.

Industry needs to benefit from Extended Enterprise techniques by involving all people throughout the product life cycle (suppliers, customers, design, production, servicing, *etc.*) enabling them to provide their product knowledge to enhance product development and support. This knowledge needs to be saved and managed; loss of this knowledge results in increased costs, longer time-to-market, reduced quality of products and services. By improving products and customer support manufacturing companies will be more competitive, and employment will increase.

Industrial companies need to shift towards the use of EE technologies and knowledge management for customer/product support. This paradigm implies a quite new scenario: knowledge capturing and sharing, new forms of interrelationship between companies and persons, *etc.*

Companies need to be able to extend their own enterprises (by removing barriers of geographic location and human related problems) to encompass the customer's operations where the supplied industrial products are being used. They need to provide the expertise to support the products *in situ* (including problem solving support, and diagnostic analysis of customer feedback), just as if the company's expert was there with the customer solving the problems. This involves EE models of the technical expertise of the company in supporting their products at the customer's site.

The key idea behind the EE concept is to develop means supporting the collection of all useful knowledge throughout the EE for new and existing process and product developments, and to develop this knowledge into a means of fostering industrial innovations. Innovations are achieved by combining the ideas and feedback from all parts of the product life cycle, including customer interaction with existing products and new product ideas, customer service and field engineers, including suppliers, and including also a pooling of knowledge between multiple sites.

This new paradigm addresses an issue of significant importance to industry: the use of "e-business technologies"[6] for EE product knowledge systems permits ubiquitous human interaction, across and beyond industrial organizations, getting organizations to work better with each other (see Chaps. 4-5). This on its side produces a significant impact on competition, employment, working conditions, internal market and free circulation of goods, health, environment, transport, innovation, and long-term sustainable growth.

The main approach is to focus on product knowledge which comes from the agents in the Extended Enterprise (EE) involved in the development, support and use of products. These agents may come both from the external EE (suppliers, customers, *etc.*) and internal EE (involved functional areas of the company) in the form of tacit or informal knowledge generated by employees. It represents the next evolution of product information systems, taking standards and practices forward to support co-operative working and partnerships.

The main business benefits arising from this new paradigm are:

- Reduction of product innovation cycle-time
- Reduction of time and efforts for solving product/process problems
- Improvement of process efficiency and reduction of wastes

The means needed to support the Extended Enterprise paradigm are:

- Means of stimulating the creation of innovative ideas and collecting them from people involved with the products and processes. Specifically to increase the number of innovative suggestions for new concepts and reducing the time ratio for new designs in the companies
- New ways of processing these ideas and storing them into a structured knowledge repository to ensure that all useful knowledge (innovative information) is saved
- Means of analyzing innovative knowledge to determine which is useful, and which is not. That is, to enable the viability of ideas to be assessed
- Means of delivering the innovative ideas to product and process designers for maximum effect

---

[6] The term "e-business" encompasses a variety of ICT tools aiming to enable many different business processes among companies through Web based ICT applications.

2.3.1 *Creativity in the Extended Enterprise*

Creativity is defined as the "Ability to produce something new through imaginative skill, whether a new solution to a problem, a new method or device, or a new artistic object or form" (MacHale 2002).

This can actually be done on an individual basis but it is not easy. In fact people are very creative in childhood but most of them bury their creativity in time under layers of rules and norms, counter-creative education, boring tasks and corporative restrictions as well as a growing (with age) fear to fail.

Most of the scholars agree that team work fosters creativity by adding extra value to the simple addition of the individual skills of the team members. Moreover most of the existing tools for creativity though they could be used in an individual pattern (but not all) are actually intended for team working.

The real challenge on a collaborative environment (Sorli and Stokic 2008) through the EE is how to "re-invent" these tools in order to enable them to be used within the new virtual working frame, create new ones and eventually integrate all them (see Chap. 4). A very important drawback is that virtual environments fail to create the warm, human, freewheeling atmosphere necessary in any "creative process."

2.3.2 *Managing Product/Process Knowledge in the Concurrent/Simultaneous Enterprise Environment*

On this framework industry in the twenty-first century has to face these challenges by using techniques to deal with aspects such as:

- *Extended Enterprise.* As already discussed, enterprises are surpassing physical boundaries and are establishing durable links with other companies — engineering, sub-contractors, providers — but are mostly at a loss on how to deal with customers at both ends of the chain. The customer is clearly a very relevant actor at the conceptual phase of the product life where the designer has to understand customer's needs and feelings as well as at the other end of it when the extended product has to live together with the user along its operating live.
- *Concurrent enterprise.* As the idea of EE refers to a longer time frame, concurrent enterprising focuses more on the specific relationship among companies to set up new operations, new product development and launch, marketing activities covering a wider range than only the physical product by itself (extended product), and others.
- *Extended product.* The concept of product is rapidly changing from the physical tangible product to the idea of providing an overall function, for example

Rank Xerox[7] became a provider of document services from being a manufacturer of photocopier machines through following a business change of paradigm in the decade from 1980 to 1990 (Stim 2006). This new focus may imply an overall strategic decision on changing the business orientation but at the very least it represents a plus of intangible assets related to fulfilling requirements, fitting the right product to the right needs, servicing the product and maintaining it through its life, empowering the user to get the best from it, and lastly facilitating the product retrieval and eventual replacement in an environmental friendly manner.

- *Support of ICT.* Besides some psychosocial changes, the technical challenge is related to the massive use and incorporation in industry of the new ICT tools and Web based technologies. There is a strong human implication in the users about getting used to the new technologies and changing the way the work has to be performed.

From this basis, the new trends should be to extend the e-working[8] systems to the whole life cycle of the extended product. In such a way, new working methods will be capable of supporting the Extended Enterprise to monitor and capture knowledge from the "extended product" throughout its life cycle. This will cover from the conception of the product/service to its disposal and back to "re-incarnation," that is to say, launching improved new extended products based on the knowledge collected from the existing ones.

As it has been mentioned above, that knowledge useful to design engineers comes in many forms and useful knowledge can come from many sources inside and outside the company. A common need amongst companies is for them to be enabled to acquire and process this knowledge so that a greater, richer, centralized knowledge and information repository is available to produce better designs, faster, with greater innovation, and with less re-inventing of the wheel. The most important needs of industrial companies with regard to design are to get good products to the market place quicker, and to reduce costs related to design.

2.4 Innovation in New Product Design

Nowadays, high rates of innovation and dramatically reduced product development lead and cycle times have been shaking both practitioners and researchers of product development management. An array of ideas have been introduced under various labels: "cross-functional teams," "design for manufacturability," "concept to customer," "computer aided engineering," "black-box engineering," "platforms," "networked development," and "knowledge management" are just exam-

[7] http://www.xerox.com

[8] Same as mentioned before for "e-business", "e-work systems" stands for the ICT tools to perform different working operations

ples of such labels some of which are discussed in several parts of this book. Such concepts have created new challenges to the organization and management of the technical functions in the firms.

Product development plays an increasingly important role in the competitiveness of the companies basically through introduction of new technologies and product customization. Therefore the product development and engineering functions have an active role to play and must step out of their traditional place as a somewhat isolated expert organization. Thus, the product development organization is more directly exposed to the competitive forces facing the business and more directly involved in the strategic development process in the firm. For that reason, the product development function will continue to attract more attention by management. The traditional boundaries of the function have changed beyond recognition becoming increasingly complex and new forms of relations and direct integration of functions have been developed.

Strong and continuous efforts have been made to reduce time to market, to implement cross-functional teams, and to support project leaders. Co-development with suppliers and extended industrial networks are on the agenda of many companies. During the same period, a strong development of the "soft" area with ICT both as products and integrated with traditional products has also occurred, accompanied by the development of ICT infrastructures for product development.

Innovation in the design of new products is one of the most critical aspects for enterprises. It is really a difficult job to innovate in an industrial environment characterized in general for the urgency, the scarce resources and within a managers' culture greatly limiting creativity. Furthermore, as has been discussed in Sect. 2.2, there exist several barriers to creativity. Likewise, one can often hear "killing phrases", expressions such as: *enough of that nonsense, it is a waste of time, do not come telling tales to me...*

In consequence, another very important issue in any new product design and development is the manufacturers' capacity to add innovation in their new products and designs. The relentless race to develop new, higher quality products and simultaneously reduce time to market and reduce product cost is a major challenge for all companies, especially for small and medium sized enterprises (SME). Notwithstanding, actually there is a lot of knowledge throughout the company that is very difficult to reuse in practice: it is in old forgotten drawings, it is in the brains of employees and may be spotted in old experiences from which the "lessons learnt" have not actually been learnt.

And, on the other hand, many authors agree that innovation ability is one of the most important competitive keys in the current enterprise because to innovate implies advantages like:

- To increase market share
- To enlarge markets and open new ones
- To overpass and take advantage over the competition

- To introduce specific features in the product making a differential from the competition
- To reduce costs

The main difficulty is then to balance these two aspects apparently so opposite: difficulty to innovate and the need to be innovative.

It seems obvious that innovation cannot be left in the hands of any "illuminated inventor's initiative." In real life there are very few people really able to invent (not too practical people by the way) and furthermore just a few companies can afford to contract any of them.

Experience up to now indicates that innovation is difficult to achieve and usually arises from the ideas that some especially brilliant persons are able to generate. However, Genrich Altshuller (Altshuller 1988, 1996, 1997) the father of TRIZ methodology (see Sect. 1.4.4), and his followers concluded that it is possible to systematize innovation. But given there are few really creative people and these persons function in an intuitive manner, some tools to "systematize the innovation" are needed. People have to be provided with tools helping them to generate creative and original ideas which build up the basis for the innovation process. This is the origin of TRIZ, a methodology helping to resolve any kind of "inventive problem"[9] providing innovative design concepts and fostering innovation.

2.4.1 Understanding the Meaning of Innovation

Though previously mentioned, it is important to agree on a common understanding about some of the expressions used in this book.

It is intuitively known what "an invention" is. Anyway it can be defined as the ability to develop a new idea either creative or different, aiming to improve a specific situation in any field: product, process, service…

However, "innovation" is only achieved when this idea ("invention") is successfully implemented. In manufacturing companies this is usually referred to as "industrialization of the idea."

Finally the level of creativity should be described as the difference that makes an idea be considered as an invention.

To complete these definitions the following contribution from Altshuller (1988) is quite relevant. By conducting an exhaustive patents research, he concluded that the technical solutions involved in the patents had a wide range of creative content. Based on that approach, he established a semi-quantitative scale to "classify" creativity. This scale, not quite scientific but very useful, is shown in Fig. 2.4.

---

[9] The concept of "inventive problem" within TRIZ philosophy means the kind of problems for which there isn't any known solution. These problems can actually foster innovation.

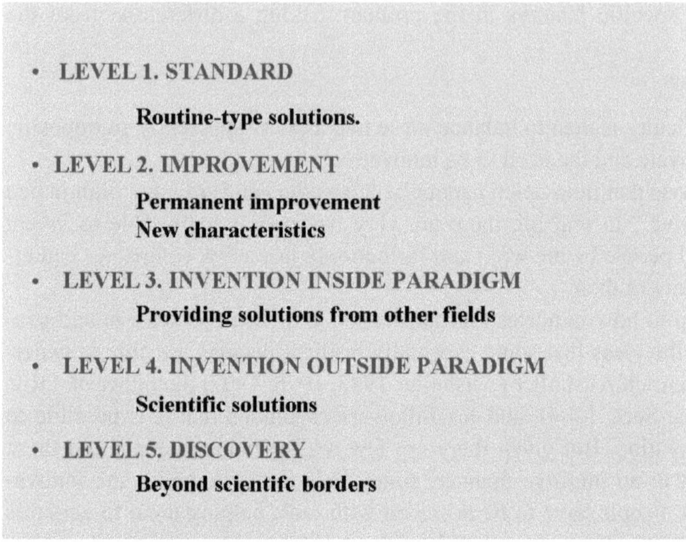

Fig. 2.4 Levels of innovation

- *Level 1 Standard.* This level represents solutions of routine-type design problems, obtained using methods well known within the particular field of expertise. In solutions at this level the existing system is not changed, although particular features may be enhanced or strengthened.
- *Level 2 Improvement.* These are solutions that, while basically leaving the existing system unchanged, do involve new features and lead to definite improvements. Inventions at this level are achieved by methods well know within the same industry.
- *Level 3 Invention inside paradigm.* These are those that constitute an essential improvement of an existing system. Level 3 inventions usually involve technology known in other industries but not widely known within the industry in which the inventive problem arose. Solutions to level 3 problems thus create paradigm shifts within their industries; they are found outside the range of accepted ideas and principles of that industry.
- *Level 4 Invention outside paradigm.* Level 4 inventions are characterized by solutions found, in Altshuller's words, "not in technology but in science," through the utilization of previously little-known physical effects and phenomena.
- *Level 5 Discovery.* These solutions are usually beyond the limits of contemporary scientific knowledge. The solution requires the discovery of some new phenomenon that is then applied to the "inventive problem." Level 5 inventions usually lead to the creation of wholly new systems and industries. Lasers, aircrafts, and computers are good examples here.

Obviously enough, words "invention" and "creativity" can only be considered from level 3 upwards.

2.4.2 Industrial Design

The increasing technical complexity of new products, besides a reduced life in the market – in addition to other characteristics as can be seen in Table 2.2 – puts into question the traditional old design process (based on independent and sequential phases) and forces the companies to choose new alternatives.

Juran, a renowned total quality "gurú", in his classic book "Quality Control Handbook" (Juran 1962; Juran *et al.* 1983) made a still valid clear distinction between *traditional* products and *modern* products, as can be seen in Table 2.2.

Table 2.2 Comparison between traditional and modern products (Juran, Quality Handbook)

	TPYE OF PRODUCT	
CHARACTERISTICS	TRADITIONAL	MODERN
Simplicity	Simple, static	Complex, dynamic
Accuracy	Low	High
Interchangeability needs	Limited	Extensive
Consumable or durable	Basically consumable	Basically durable
Using environment	Natural	Artificial
Product understanding by the user	High	Low
Importance for human health, safety and life continuity	Rarely important	Usually important
Life cycle cost for the user	Equal to purchasing price	Bigger than purchasing price
Life of a new design	Long: decades or even centuries	Short: less than a decade
Scientific base of design	Empirical in large measure	Scientist in large measure
Reliability, maintainability, availability issues	Scarce	Quantitative
Production volume	Usually low	Usually high
Usual causes of failure in service	Manufacturing failures	Design failures

Traditional products can be shoes, garden hardware, bread, or aspirins. Modern products are printed circuit boards, computers, aircrafts *etc.* The first telephones and cars were traditional in their simplicity, but now they are modern in their complexity.

Following this idea, two processes are needed for launching new products: *traditional processes* for traditional products and *new processes* for new modern products. Traditional methods are inadequate for modern products. They have been tried but they fail. Besides, use of modern methods for traditional products appears to be uneconomical.

So, following Juran, companies with traditional products should not have too many problems with their current design processes whereas those incorporating modern products should revise them.

However, by now, this may be considered as not completely true. Both continuous improvement and innovation have room in either of the two processes and furthermore can be applied both to the product development and to the correspondent production processes.

In general, the most outstanding and urgent needs industrial manufacturing companies must cope with are:

1. *Reduce time to market.* Most of the design activities are still based on "pen and paper" work. However, providing the right knowledge at the right time can increase competitiveness by reducing project timescales, avoiding repeating mistakes, and enabling solutions to be generated faster.
2. *Reduce costs arising from design.* Design costs amount in general to a significant share of the company overall costs and can be reduced by improved working, better access to information, less time to look for information or search for ideas, and more people focused on solving design problems. Companies need to re-use good design ideas from the past designs (less re-inventions).
3. *Make better products.* Innovation is a critical factor in the success of industrial companies. They need to develop innovative products and solutions and simultaneously reduce the design time, comply with regulations (national and supranational), take into consideration recycle-ability and energy consumption issues, *etc.* This is very difficult for companies, mainly SME, which are in general less able to translate ideas or knowledge into innovations.
4. *Increase involvement in design.* By increasing the number of people (with a structured system) involved in providing design inputs and knowledge, it is possible to enrich the design process, and also improve motivation by involving people. The knowledge of the product/process is distributed across the whole company. This know-how represents an essential resource for successful competition in the market and should therefore be preserved and used as efficiently as possible. There is a risk that this knowledge is lost when key persons and engineers leave the company.
5. *Right first time.* Companies need to get it right first time. Reworking designs and recalling products are every company's nightmare, particularly so for SME.
6. *Reduce maintenance.* By making better products and incorporating the knowledge of maintenance engineers (feedback from maintenance to design, which is not often present in a structured way), maintenance costs can be reduced.

There exist means providing practical methods for capturing, storing, reusing, and developing knowledge into innovative and quality designs. New ICT-based approaches (see Chaps. 4 and 5) of processing knowledge are required to manage all the diverse forms of knowledge that design engineers are exposed to. It will help them to make best use of the extended knowledge resource of companies, increase the development rates of innovative ranges of products/solutions, reduce design time and costs, increase customer satisfaction, improve the process of new products/processes development, and achieve an overall business success.

2.5 Risks in Innovating in New Product

2.5.1 Main Difficulties for Innovation

Innovation is a risky business as well as design of new products, as will be discussed in Chap. 3 (Sect. 3.1.1). Combination of both – "*Innovation in new product design*" – should very likely increase the risk level. In consequence, there are some aspects that have to be considered in order to minimize these risks, which are going to be discussed in this section.

In the product design process, people involved within the product life cycle and the production processes have to be encouraged to generate innovation. Team working between people from different sites (and working off-site) and between organizations, customers, and suppliers along the Extended Enterprise (EE) also needs to be encouraged. *Systemic Innovation* in Chap. 3 (Sect. 3.4) and *Open Innovation* (Chap. 6, Sect. 6.4) will discuss these issues in more detail.

The accelerated pace of technological development continuously increases time and market pressures on manufacturers' capacity to innovate new products and designs and to develop the manufacturing processes that produce these products. As said before, there is a real need for companies to develop new products with higher quality, reducing time to market, product cost and improving quality, but many companies lack the financial capacity either to invest in the latest technology as it reaches the market or to hire specialists to integrate new methodologies and systematically to improve their products.

Many companies would achieve the required corporate breadth-of-experience to improve their products and improve their processes if they could only make best use of their knowledge resources internally and in partnership with their suppliers and customers. Stimulation of "innovation" is a means by which these knowledge resources could be channeled.

Major difficulties for innovation are related to three main topics (which are addressed throughout the book):

1. *Intangibility of the inventive knowledge.* The inventive capacity is usually considered more as an inherent property of the genius than something that may be learnt. Intangibility makes the inventive knowledge difficult to accumulate and transfer. Emerging theories say that the capacity for innovation observed in some inventors is no more than an instinctively applied methodology for abstraction, which gives sense to the words "inventive knowledge" (or "innovative knowledge"), defined as "the knowledge necessary for finding solutions at any abstraction level." Therefore intangibility will be overcome by establishing

rules, methodologies, and tools first for abstraction (general rules or solutions) and then for concretion (applying the general rules and solutions to the specific situation) of problems, allowing accumulation of them and their solutions in a hierarchical database with the abstraction level as hierarchy separator.

2. *Individualization of the innovation process.* Investigations performed during the last 20 years have demonstrated that innovation is better achieved by working as a team. In the first conceptualization steps the working teams should include the best experts in several fields available world-wide which becomes quite impracticable in the current stressed and time limited working environments for most industrial companies (mainly for SME). Due to this problem, innovative thinking is hardly tried by individuals on their own and the results are generally poor (geniuses are not so frequent).
3. *Information overload.* (Goldratt 1990) Nowadays there is a huge amount of information coming from many diverse sources but finding the needed information and knowledge in the right moment is becoming a real problem. Traditional methods like looking for and directly contacting the right person for the right knowledge are becoming almost impossible in the increasingly isolated and time driven working environments of today. Team working among stakeholders in the product value chain is then the only reasonable way of overcoming this difficulty.

Such problems could be minimized by employing innovation methodologies during the development process and incorporating tools to support innovation along the way. However, even when enterprises try to incorporate new methodologies, many problems appear due to human – and methodology – specific factors.

Human factors include problems of encouraging and convincing people to use new and innovative methodologies. It is noted that new methodologies, however enthusiastically received, are frequently discarded in favor of familiar methods shortly after they are taught and personnel are trained. Implementation of new methodologies is also frequently inefficient in time-management terms due to complexity, dependence on worker experience and interpretation, as well as processing of results.

Methodology factors, *e.g.,* available engineering methodologies, are frequently theory-overloaded and do not integrate well with one another, if at all. In the chain of methodologies there is lack of transparency in planning, cost, and technological and quality data.

Key aspect to shift from the current ways of working to the "New Paradigm" as it is going to be presented in Chap. 3 (Sect. 3.1), is the Extended Enterprise (EE) concept previously addressed in the book. The main challenges to be faced under this paradigm are:

- Developing practical means of developing ideas into innovations in products and processes. This will involve taking what is currently available and producing methods of rapidly taking many creative ideas, and assisting people to work together in a structured manner to develop these ideas into innovations.
- Capturing and structuring of innovative ideas over EE in such a way that they can be best used for product/process innovation; this is typical "difficult to structure knowledge" which asks for high level "innovation" meta classification – on one hand the structure must not restrict creativity of the people, on the other they must be structured in such a way as to be easy to assess and reuse.
- Providing means for team development of innovative ideas over EE is a tough challenge and asks for generic approach for development of ontologies applicable in the context of specific products/processes.

Specific innovations that have to be incorporated to the company's culture are:

- *Stimulating the creation of ideas* about products and processes throughout the EE, empowering all people coming into contact with the products or processes to provide their thoughts on improvements and original ideas
- *Interactive solution to be able to take basic ideas* and develop them (by *collective working* throughout the EE), into product and process design innovations
- *Development of diverse ideas from multiple sources* into workable innovative designs (for industrial products and processes)
- *Assessment of innovative ideas* to analyze their likely success, and thereby evaluate the viability of ideas/designs
- *Development of specific ontologies* needed to enable efficient exchange of ideas between different experts/actors within an EE
- *Combination of methods* for creating innovative ideas with "classical" methods for collection of knowledge on products/processes and problems
- *Development of a combination of repositories* with innovative ideas, products-processes knowledge, and information/knowledge on problems, and/or improvement potentials
- *Fostering new forms of organizational learning* within the EE by collecting and storing innovative ideas and making them available over long time period

Since the basis is sharing innovative ideas from different actors within EE the involvement of the end-user is critical. This provides a good basis for efficient specification and testing of methods and tools, taking into account critical human related aspects.

2.5.2 Risk Management

2.5.2.1 Risks

Definition of risk varies widely mostly depending on the context in which it is used. In the case of innovative new products development, risk can be considered as "any event that provokes undesirable effects in the process which will finally result in economical losses for the company."

It is relevant here to consider a distinction between "uncertainty" and "risk" (Knight 1921) in the sense that a risk can to some extent be assessed (measured) while an uncertainty is almost impossible to measure.

Hubbard (2007) proposes a more detailed definition of both terms and how to measure them as:

> **Uncertainty**: the lack of complete certainty, that is, the existence of more than one possibility. The "true" outcome/state/result/value is not known.
> **Measurement of uncertainty**: a set of probabilities assigned to a set of possibilities. Example: "There is a 60% chance this market will double in five years."
> **Risk**: a state of uncertainty where some of the possibilities involve a loss, catastrophe, or other undesirable outcome.
> **Measurement of risk**: a set of possibilities each with quantified probabilities and quantified losses. Example: "There is a 40% chance the proposed oil well will be dry with a loss of $12 million in exploratory drilling costs."
>
> D. Hubbard

The final idea (according to Hubbard) is that uncertainty may exist without risk (uncertainty in a weather forecast is not risky for office work) but risk always implies uncertainty (risk in navigation may come from uncertainty in weather forecast).

2.5.2.2 Fundamentals of Risk Management

Risk management. This is a key aspect of project management. In any project a risk analysis should be performed beforehand. Furthermore, in a process aiming at innovation in new products design and development, risk management becomes of an utmost importance. Its results will become the input for a decision making "go/not go" gate.

Risk management builds upon three legs that are discussed in next sections:

- *Risk analysis*
- *Contingency plan*
- *Decision making*

Risk analysis process should be a continuous activity throughout the project combined with the decision making gates and a contingency plan intended to mitigate the potential negative impacts of the risk occurrence. The combination of the three constitutes the risk management.

Risk analysis. Risk analysis is an important but highly difficult activity in each project but in contrast, it pays back well since once the risks have been identified and assessed it becomes easier to implement measures to prevent their occurrence or/and mitigate the resultant effects.

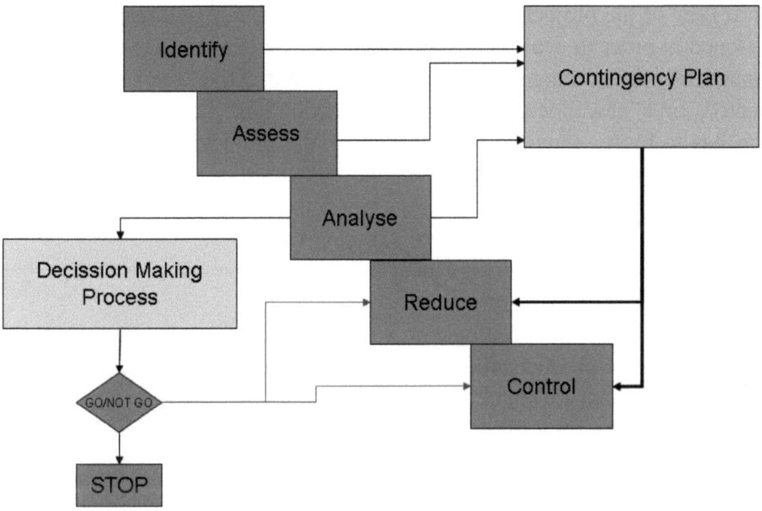

Fig. 2.5 Risk management process

As can be seen in Fig. 2.5 adapted from Roy (2003), the process follows five steps selectively concentrating on those risks with higher probability of occurrence or more serious impact on the process:

1. Identifying the risk
2. Assessing the risk: evaluating the impact and probability of occurrence
3. Analyzing the risk: understanding the process for which the risk may show up
4. Reducing the risk: setting up any feasible means that could prevent the risk occurrence or its impact on the process
5. Controlling the risk: monitoring the process trying to prevent the risk before it actually comes out or to trigger countermeasures once it shows up

These countermeasures (5) should have previously been identified and defined in the contingency plan; meanwhile any of the steps 2, 3 or 4 should derive to a decision making gate.

Contingency plan. A contingency plan must implement all possible preventive actions that should prevent risks occurrence but must also contain alternative plans to be launched upon the appearance of identified risks as well as how to proceed in the case of appearance of non-identified risks. It should follow the form of a flow chart showing the alternatives to be chosen in case of deviations from the original plan ("if…then") with as many branches as possible. A flowchart of the kind can be prepared by using tools such the "Process Decision Program Chart (PDPC)" in which the process flow is represented describing all possible situations with the required actions to correct deviations (see Sect. 1.4.9).

Decision making. Decision making gates have to be pre-defined in time related to the standard flow of the project but also an agile decision making mechanism has to be developed in order to deal with risks appearance. Whenever a risk is becoming unmanageable, the contingency measures show ineffective, the risk is higher than expected, or a new unexpected risk is showing up, the decision making mechanism has to be triggered, and a decision on the project continuation or on changes to be implemented in it has to be made.

2.5.2.3 Identifying and Assessing Risk

Risks are evaluated by means of allocating weight to them in a numeric form. In general a risk equals the product of its probability of occurrence times the value of the impact on the process (*i.e.,* potential losses):

$$\text{Risk evaluation} = \text{Occurrence} * \text{Impact}. \tag{2.1}$$

The result of such an equation provides a priority number enabling the analyzer to concentrate its work on those with higher priority number (third step: analyze). In this equation, occurrence means the statistical probability of the appearance of the *cause* (event) that will trigger the process resulting in an undesirable *effect.* A serious drawback lies in the difficulty of evaluating both concepts; there are several tools that allow one to estimate them.

Tools to calculate the probability of occurrence can be grouped under the general umbrella of Probabilistic Risk Assessment (PRA studies):

- *Fault Tree Analysis (FTA).* FTA (Mobley 1999) by using logical gates (and, or, not) represents in the form of a tree the sequence and combination of failures that chain up to the final effect.
- *Human Reliability Analysis (HRA).* Methods for evaluating and modeling human errors (Pekka 2000). Detailed breakdown of human tasks allows the assignation of probability of failures to "human systems" in the same way as FTA works for technical systems. This method is most related to processes with heavy safety implications (Gertman and Blackman 2001).
- *Common-Cause Failure Analysis (CCF).* Methods for evaluating the effect of inter-system and intra-system dependencies which may tend to cause simultaneous failures so increasing the overall top effect.

Following any (or a combination) of these techniques, the effect or final result of the failure is traced down to the individual causes at bottom level for which in many cases there are statistical data of probability of failures: information from databases catalogues from the providers, bath-tube graphs, reliability analysis, *etc.* Tree hierarchical structures allow calculating upwards of the probability of occurrence of the effect.

Impact evaluation. The impact is related to the effect caused by the failure and has to be considered as the "cost" of the undesired result. It can be estimated in monetary units as the expected cost of the failure for the company allowing sorting by their level of magnitude. In cost engineering (Roy 2003) economical evaluation of risks helps to estimate better the overall costs of the development and to decide if they are assumable by the company.

Nevertheless it should not be forgotten that the final goal of this evaluation is to sort the risks by order of priority, so in practical terms (as the following tools show) impacts can be evaluated by means of assigning them just a neutral figure which represents the "value" of their effect. Those which may eventually cause damage to persons or properties have to be taken up to the higher level of priority independently on how the cost of a human being could be evaluated (*i.e.*, by insurance companies).

The following tools support both identification and assessment of the potential risks.

Failure Mode and Effect Analysis (FMEA). In which an experienced team develops an in deep analysis of all possible failure modes and assigns to each of them, three factors (see Sect. 1.4.5):

- Probability of occurrence of the cause of failure (O)
- Severity of the effect of the failure (S)
- Probability of the failure being detected before the occurrence of the effect (D)

$$O*S*D = RPN \tag{2.2}$$

The product of the three factors (D in inverse mode) provides the "risk priority number" (RPN) in order to sort out the list of failures according to their expected impact. The three factors, independently of how are they calculated, have to be conversed to a common rank in order to have a consistent result.

Preliminary Hazard Analysis. It is similar to the FMEA in its method but focuses in greater detail on the hazardous incidents related to industries potentially dangerous for the human being and/or the environment.

Anticipatory Failure Determination (AFD). AFD is an interesting approach arising from TRIZ methodology (see Sect. 1.4.4) which systematically employs ARIZ-TRIZ algorithm to "invent" any possible mode of failure in the system. Similar to FMEA, it helps to develop an exhaustive list of failure modes and then to evaluate its probability of occurrence by means of logically analyzing the combination of events that may be produced in determined circumstances. AFD does not utilize numeric calculations (like FMEA) nor logical sequential chains (as FTA) but just rational inventive thinking (TRIZ).

2.5.3 *The Human Factor in Risk*

It can be clearly seen that the evaluation of both factors of the risk equation – mostly the impact – is highly subjective though the contribution of the team and the use of the above-mentioned "neutral techniques" tend to minimize the subjective influence. Affect, emotion, personal perception, "gut feeling," and other factors (Slovic *et al.* 2004) form part of what Epstein (1994) named "Experiential System" which has been built up over years of human evolution largely based on affections.

The experiential system will then have a preeminent role in the twofold decision making process involved in risk management: decision on the assignation of priority to the risk and decision on how to manage it through the contingency plan and the decision-making gates process.

Another important aspect to be taken into account is the fact that the "objective reality" coexists with the "subjective reality." On one side, each person has or may have on his mind a different perception of the reality and on the other the way reality is described by someone will actually built up a new one in the minds of the hearers.

Some interesting facts to highlight from Slovic's work (Slovic *et al.* 2004) on the experimental system are:

- People base their evaluations not only on what they think of it – which can have a sound technical and scientific basis – but also on how they feel about it. Previous experiences, misjudgements, and prejudices may bias the decision to the direction of the feelings.
- The way data are provided has an utterly influence on the decision. Percentage, probability, and statistical figures have a quite different meaning to different evaluators since rough absolute figures are difficult to compare.
- Insensibility to probability. In determined circumstances people tend to dismiss probability ranges, for instance playing the lottery expecting to achieve huge winnings against the very low probabilities of such a thing happening.

It can be concluded that "experiential" decision making is important as it is based on intuition, affections, feelings, and other personal aspects that may help to make a good decision more quickly. On the other hand, rational, analytical decision making based on data and facts is usually more time consuming but the results should be sounder and more reliable.

The important point to highlight is the fact that no matter how accurate the information and available knowledge might be, there always will be the variable influence of the human factor in the form of feelings, prejudices, intuition, and so.

Being aware of this and, depending on the required agility and importance of the decision to be made, a good balance between experiential and rational systems should be achieved.

2.5.4 Risks in Innovation

The following risks may be the most serious possible risks that can be present in the development and launching of any innovation in product. These can be considered as the upper level risks arising from different combinations of smaller risks at lower levels:

- Innovation is not accepted by the market
- Costs increase out of control swallowing expected benefits
- Development time exceeds all forecasts
- Innovation turns out to be a bad solution
- Product failures show up once the product is on the market

The three domains economic, technologic and societal have to be well balanced in order to mitigate the revolution caused by a breakthrough change in the market.

Innovation may not be accepted by the market, usually because it comes out at the wrong moment for some reason: it is too revolutionary for the time, it is blocked by the economic situation, not in tune with the societal culture and time, *etc.*

The De Lorean case is a paradigm in that sense but there are more examples such as Sony's Beta video system, the Citröen SM model launched in 1970 that failed completely in the market, Apple's electronic agenda "Newton" launched in 1993 with the right technology (incorporated later in the modern PDAs) which also failed in the market until being discarded in 1998. Nowadays it can also be seen in the market penetration problems of the hybrid vehicles, being very reluctantly accepted by the customers.

A comprehensive analysis of the market trends, the use of prospective tools (Delphi, expert's consultation, surveys, *etc.*) combined with clinics and pilot testing with friend-users can help to minimize this risk. QFD (see Sect. 1.4.3) supporting the identification and whole analysis of customers' needs and requirements also helps one to understand better how to fit the products to the market but fails in predicting if the innovative solutions adapted to fulfill some requirements would be too "scaring" for the market.

Another interesting tool for this purpose is "Direct Evolution" (one of the TRIZ applications – see Sect. 1.4.4) that provides a good methodology to analyze the historic evolution of technological systems and predict their future trends.

Risks of uncontrolled increase of costs, excessive development time, or inappropriate solutions can always occur but their appearance is fairly reduced with a solid and well managed development process as that described in Chap. 3.

The case of failures showing up once the product is on the market is the most dangerous for any industrial company due to the costs not only in monetary units: warranty, reposition, post sales service and maintenance, *etc.* but far more important, the intangible costs of loss of brand image, and customer dissatisfaction with the very dangerous (for the company) "mouth to ear" propagation of the complaints that is estimated to multiply by more than five the number of potential customers that will be eventually lost by the brand. The proposed new development process also pays special attention to this specific problem.

From a study realized in five small and creative companies in the UK in 2008 (Jerrard *et al.* 2008) the following risks categories were identified inside and outside the company:

- Financial: operational finance, access to working capital, pricing
- Personal: personal finance, family circumstances
- Intellectual Property: developing and protecting ideas, research needs
- Regulatory compliance: policy changes, safety issues, new standards
- Markets: competition, consumer/customer response
- Technical: manufacturing processes, new technologies, components
- Partnerships/collaborations: networks, cross-functional teams, formal/informal partnerships, *e.g.,* suppliers, specialist input, distribution networks
- Organizational: capacity, skill, support/commitment to NPD

R. Jerrard

Summarizing this list, the key points related to risk in new process development (NPD) are related to:

- The way risk is assessed in decision making
- Communication between the design team and the decision makers
- Acceptance of the risk as inherent to the creativity process
- Balance between the risk assumed by designers and the risk accepted by consumers (are they willing to assume any risk at all?)

2.5.5 Minimizing Risk in Product/Process Development

The overall innovation process, described in Chap. 3 as the "*New Paradigm in Product/Process Development,*" is focusing upon the minimization of these risks. This solid and coherent development process allows one to take into account most of these risks from the early stages of the process, so minimizing the risk occurrence.

Categories of risks defined by Jerrard above (Jerrard *et al.* 2008) can be summarized by grouping them into four categories:

- *Financial.* Company's financial resources and Intellectual Property Rights (IPR) issues
- *Human related.* Personal situation, personal behavior (teams, partnerships, networks, *etc.*) and organizational
- *Technical.* Including regulatory aspects
- *Marketing*

Save for financial aspects, the rest of the categories are analyzed and dealt with in Chap. 3 as well as mentioned again in several points in the other chapters. Financial aspects are clearly out of the scope of the book since it is another domain; nevertheless, a well controlled process such as the one to be described next has the advantages of being quite accurate in the costs estimations and providing high confidence to potential investors and financial entities.

Chapter 3
Product/Process Development Process for the Twenty-first Century

Abstract As discussed in previous chapters, new requirements from the twenty-first century force manufacturing companies to undergo a revolution in the way they are used to doing things. One of the biggest changes is related to the process for new products and process development which is, as previously discussed, most important in terms of long range competitiveness in the market. Main drivers of the new process and the change process are presented and described in this chapter. These drivers are: new paradigm, the new working model, systemic innovation, and the 3 Cs model: customer driven, concurrent engineering and collaborative work. Influence of information and communication technologies (ICT)-based methodologies and tools in the new working modes and the Collaborative Working Environments (CWE) as a specific sub-system of the ICT are just drafted and will be analyzed in further detail in Chaps. 4 and 5. Finally, a specific section deals with systemic innovation in new product development.

3.1 New Paradigm in Product/Process Development

This section will realize an overview of the current trends on product/process development within the new industrial situation and the foreseeable evolution in the near future, what can be considered as a *new paradigm.*

3.1.1 Launching a New Product

As has been discussed before, “new products” is a broad expression that may range from minor lifting to a radical new product or invention. As a matter of fact, nowadays there are quite few really “new products” (we may call them “inventions") since new developments are mostly based on existing ones. Most of the inventions (patents) are new solutions to old problems with the same old products. Generally speaking, this approach may be valid for manufactured goods but even in the field of information technology after the explosion of the "World Wide Web" (www), most of the innovations or new products actually follow the same schema.

The real level of change varies depending on several factors and the most important is driven by customer perception: *if the customer feels that the product is really new, then it is new*, whatever technical changes there would actually be inside it.

Under this approach, decision on launching a new product into the market is always a rough one but sure enough it is a "must," and, if successful, it pays back very well. As a matter of fact, successful stories in industry tell us that 80% of revenues come from products developed in the last five years.

The decision on initiating a new product development process has to be soundly based on:

- *Strategic analysis* of the company's situation in order to prioritize these product families presenting the better revenue expectations for the company
- *Market knowledge* to develop a product aiming to a specific target and well fitting with this group's needs and requirements
- *Foreseeable future trends* of the market needs

A good knowledge of these factors is unavoidable if one wishes to develop a new product with the highest success probabilities, but besides these factors, there are still others that the entrepreneur has to be quite aware of:

- *Need and timing.* Both factors are of capital importance. If there is not any specific need detected in the market, a new product ("invention") may just fail in spite of being a really brilliant idea. History of technology is full of examples of good products/inventions that have failed mainly (among other considerations) for being in advance of their time. Not pretending to enter into a discussion about the multiple causes that may bring a product to failure in the market, there are some good examples as previously discussed in Chap. 2 (Sect. 2.5.4). All of them were in general good products, highly innovative in concept as well as in technical terms but they failed in the market for several reasons, the most important being probably that the market was not prepared to adopt the innovations they presented.

  It is true that the need may be somehow generated in the market arena but only to some extent. Introducing a new invention in the market is usually a matter of timing:

  - The product has to show up just at the moment when the need is starting to arise
  - The market needs time to adopt it and get adapted to it

- *Marketing strategy.* Closely related to the previous point, it can surely help to raise and strengthen a hidden need and create a state of mind letting people become really anxious to fulfill it. Never forget that it may only be true if there is a real need to fulfill. Poor strategies aiming at the wrong target or choosing the wrong need will surely end in product failures.
- *Capital investment.* Even brilliant ideas can't survive shortsightedness of potential investors. Capital shortage may make impossible the manufacturing or commercializing of good ideas. In the current market, an idea should not only be unique but it has to reach the market within the buying power of the target user.
- *Luck.* Finally there is an intangible factor which is nevertheless really unavoidable. There is a real need for some doses of good luck, but one must be aware that luck is not just fortune, but a combination of aspects such as hard work, good sense, deep conviction, and persistence.

In summary, there are so many aspects taking part in the product development process that sometimes it is amazing how new products are still being produced everyday, everywhere.

3.1.2 Lead Time

In current times it is quite obvious that one of the most decisive success factors for industrial firms is the so called "lead time" or launching time for new products.

"Lead time" or "time to market" has been generally admitted to be one of the most important keys for success in manufacturing companies. A combination of factors such as ever changing market needs and expectations, tough competition, and emerging technologies among others, challenges industrial companies to increase continuously the rate of new products to the market to cope with all these factors and requirements. As has previously been discussed, hitting the right date to come to the market ahead of the competition gives industrial companies a competitive advantage and a better market penetration rate.

Lead time actually covers from the product conception to its commercialization in the market place. It obviously varies greatly depending on the type of product. Usually there is an increasing need to reduce lead time in order to enable the company to increase the rate of product refreshing in the market as well as better adapting to the rapidly changing demand from the customers.

Another issue is how to measure "lead time" adequately. From one end, the *starting point* may not be quite clear since development of complex products *(i.e.,* cars) usually carries on at different speeds and time intervals for each of the components. Also, at the other end, *finalization* of the process can be defined as the time of delivery of the first product. Unfortunately this indicator can be cut down for the real first parts by using extraordinary short-cuts such as making special arrangements or pushing production means, *etc.,* but could not have continuity for the following production due to unsolved or unstable situations in production, procurement, logistics, *etc.* Another indicator can be the point where production has achieved its "cruise speed" which is also quite uncertain to be chosen. The last indicator that is being used is the break even point, the time when the financial balance between inversion and income is achieved.

Nevertheless, it is more important to use the concept of "*management of product development time*" rather than simply looking for a time cut-off. Under this new paradigm, companies capable of "mastering" the development time will be able to launch the product into the market while just spending the planned time and resources and at the *best* moment, meaning the exact date when the product is expected to achieve the highest and fastest market penetration. This will give back to the company a higher market share and better returns.

This new paradigm will only be achieved through the integration of three large sets of tools:

- Techniques from total quality management (TQM)
- Concurrent engineering (CE)
- Information and communication technologies (ICT)

3.1.3 Innovation

Innovation should actually be embedded in the process of development of a new product, otherwise it will not be a *new product* but a combination of old solutions upon an existing product which will hardly pay back any benefit to the manufacturer.

So the process for new product/process development can itself be considered as an innovation process on the whole. Nevertheless there are some specific phases of it in which innovation is even more important: *concept design* and *detail design* to be described when discussing the phases of the new model (see Sect. 3.2.2).

Innovation has a high level of risk; in Sect. 3.1.1 a baseline on which the process for launching new products has to be supported in order to prevent risk and/or minimize risk's impact has been discussed. Also, some of the problems that will be encountered in the innovation process have been discussed in Chap. 2 (Sect. 2.5), including human related barriers to innovation and risk management.

3.2 New Model Within the New Paradigm

This section will analyze in detail the different steps of the new model for product/process development within the new paradigm.

3.2.1 *Introduction*

The *new model* to which industry has to migrate in order to be capable of fulfilling the new requirements in the emerging scenario for the twenty-first century puts forward a framework comprising methodology, guides for action and tools based on the integration of above-mentioned fields, all underpinned by information-management technology (ICT)-based systems (Sorli and Gutiérrez 2002; Sorli *et al.* 2003). This model is to serve as an aid in developing parts in a complex, heterogeneous context featuring the participation of a number of companies in the supply chain (EE) in addition to the various internal sections of the company.

Extended Enterprise. This brings us into the new paradigm of the Extended Enterprise (EE), already explained in Chap. 2, *i.e.,* the company that goes beyond its natural limits, reaching both forwards and backwards along the value chain – to suppliers (of parts, raw materials and services, and also of technology, tools and work systems), and to customers (such as purchasers, users, maintenance people, and even those in charge of the disposal of the product).

Thus the *new model* put forward will help companies to optimize the joint application of these methodologies and their computer tools: integrated engineering as the merging of concurrent engineering (CE) plus the concept and tools of total quality management (TQM), while also adding the concept of "Extended Enterprise" including supplier integration, and drawing on the resources of information and communication tools (ICT).

Benefits of the new model. The main benefit that can be expected of the new model is a significant reduction both of time and costs which is going to be analyzed next. Research has been based on gathering data and studies setting out the benefits of the Japanese model based on concurrent engineering and on the use of certain tools and methodologies – such as QFD.

Shortening the time scale. The new process may take as little as half the time of the traditional process. Shorter lead times in product development come not from reducing the number of tasks but from making them concurrent and optimizing them.

Table 3.1 Effort percentages by stages.

	STAGES OF THE PROCESS		
COMPANY	DEFINITION	DESIGN	REDESIGN
British	17%	33%	50%
Japanese	66%	24%	10%

Table 3.1 sets out comparative data on the distribution of effort in percentage terms in the development stages for a British company and a Japanese one using CE techniques. Data are taken from a 1990 study by Prasard (1997).

This difference in effort deployment brings better product definition in the Japanese model, significantly reducing the need for redesigning. Most cases of redesign stem from a product definition that failed to take account of the potential future problems of other departments in the development process. Redesign increases cost, effort, and it always means lengthening the time span. In contrast, higher efforts in product definition are usually reflected in a considerable reduction in time, as it has been previously discussed in Sect. 1.4.3.

Reducing costs. The same above-mentioned study (Prasard 1997) shows that the use of concurrent engineering techniques brings savings of 20% in the total development cost of a new product. These objectives will more easily be achieved and improved by integration with the other techniques (TQM, EE, ICT) plus incorporating the use of tools such as TRIZ and QFD (both outlined in Sect. 1.4), as proposed in this new model for the development process.

Specifically, the use of this new model by companies will enable them to achieve these estimated goals:

- Reducing development costs by around 20% at least, broken down into:
  - Lead time reduction
  - Reduction in development hours
  - Reduction in other costs
- Reducing quality costs by 30% – in terms of the following aspects among others: warranties, repairs, repercussions of poor quality on the company's image (brand) in the market, *etc.*
- Thus resulting in a better pay-off (return on investment)

As mentioned above, these reductions are achieved thanks to putting together all departments involved into the design and development process, particularly the manufacturing department and the external suppliers in the product's value chain. This process needs to be implemented by adopting a new model of organization in which the *when, where, and how* have to be precisely defined at department, and company level, as well as how to manage relationship among all actors and perform the process management.

As will be discussed in the new model (see Sect. 3.2.2), changes must be concentrated early in the design phases where they are "cheaper" and easier to implement.

Figure 3.1 shows the differences between a traditional process and the new process as regards the quantity (amount of changes) and timing of occurrence of engineering changes.

This diagram, which was popularized by Toyota[1] (Hauser and Clausing 1988; Verma *et al.* 1996) when it began to apply QFD in the 1970s, uses real data to show how engineering changes in the traditional processes are accumulated

[1] Originally published by Toyota, this graph has been widely used in QFD related literature.

mostly in stages very close to the launching date. The reason for that is that owners of each of design stages work without any interconnection and the problems behind the changes emerge when the physical product forces people to put together and share their partial results. Moreover, with the launch date drawing very close and being unmovable, decisions are taken too hastily which leads in its turn to further changes after the product has been marketed (with great damage to brand image).

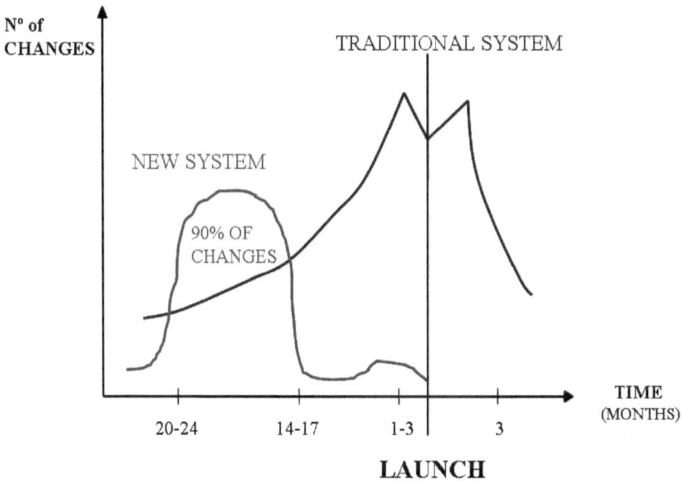

Fig. 3.1 Engineering changes. Adapted from Hauser and Clausing (1988)

However, in the new process, team work and the use of QFD along with other concurrent engineering techniques manage to concentrate the changes in very preliminary stages (at lower cost) practically eliminating them in later stages and banning them out at the production/selling stage. In addition, the overall volume of changes is reduced by nearly 70%.

3.2.2 *Stages in the New Product/Process Development Model*

3.2.2.1 General Description

Currently, the major challenge manufacturers are facing is to reduce product-development time as far as possible. This idea means minimum development time for all the parts and components. Reducing product development time is always a must but as has been said before, it is more important to be able to accomplish an accurate management of product development time.

Under this idea and in terms of the hot subject of cost reduction, the new model requires *all the departments* involved in product design and development to work closely together from the initial stages bearing constantly in mind the objectives of optimizing the added value and reducing the number of changes. Within this new approach, *all departments* means internal and out of the company teams: "Extended Enterprise."

As a result of this new context, the solution involves making a multifunctional effort, and working with common databases in the development of parts that require a great deal of information exchange – in connection with matters such as graphic data, numerical data, written documentation, but also using methodologies, techniques, specifications, *etc.* in the fields of quality (both of the development process and of the final product) and process management.

Good results can be achieved by using concurrent engineering, total quality, or EE techniques separately, but optimum results cannot be achieved without their integration underpinned by a set of ICT tools. The new model to be presented later is based on integrating all three techniques, plus software and information tools (ICT) to make it work properly.

In short, it can be said that the main characteristics of this new design and development process (new model) are based on three fundamental aspects on superimposed levels, namely:

- *Gearing company culture towards total quality*

This entails a number of changes which, to summarize from several sources (Ishikawa 1985; Groocock 1986; Juran 1993, 1995; Kume 1993; Saderra 1994; Heinezahll 1996; Zink 1997), are framed in these "ten commandments":

1. Customer orientation
2. Management leadership
3. Decisions based on analyzing the facts and the data
4. Management by processes
5. Involvement of all the staff[2]
6. Quality assurance for the product
7. Association with suppliers ("Extended Enterprise" and comakership concepts to be discussed later on)
8. Looking for results not only in the short, but in the medium and long term
9. Continuous improvement
10. Contributions to society.

- *Changes in operations and in organization structure:*
  - *Team working.* Work teams showing the usual features of present-day teams – interdepartmental and multifunctional – and then adding the new feature of also being "intercompany," *i.e.,* they also achieve the effective integration of functional areas of cooperating companies along the "Extended Enterprise."

[2] Staff from the "Extended Enterprise" covering internal and external people.

Thus the work teams are made up of staff who, in organic terms, belong to different departments of the leader company (interdepartmental) and of the cooperating companies (intercompany), and who also contribute their knowledge and experience in different disciplines (multifunctional).

– *Concurrent engineering*. Concurrent engineering is based on making the process stages overlap and concur, using the team structure mentioned above in conjunction with advanced tools and methodologies as described in Chap. 1 (Sect. 1.4): QFD, TRIZ, DoE, FMEA, Taguchi, simulation, *etc.*; underpinned by information and communication management systems and tools (ICT).

- *Information and communication technologies (ICT) based systems.* ICT issues are described and discussed in further detail in Chaps. 4 and 5. In general they require certain fundamental elements:

  – *The use of shareware and groupware systems*. These systems, drawing on a common database, facilitate a number of aspects such as sharing information, working simultaneously with the same data, off-site working, *etc.*
  – *Management by projects.* As previously pointed out, the new decentralized approach entails a significant loss of control. The new scheme implies a change in the control paradigm. Control must be focused exclusively on the project, on what can be called "project quality," and so a range of indicators must be established to enable real-time evaluation of performance to be achieved: progress, efficiency, delays, deviations, critical points, *etc.* This is what is known as management by projects. Of great interest from this aspect are the contributions of Goldratt in his books "The Goal" (Goldratt 1996) and "Critical Chain" (Goldratt 1997).
  – *The existence of interfaces for successful communication.* These interfaces are not limited solely to computer tools but rather are based, to a large extent and more importantly, on human relations.
  – *Communication channels.* Although the structure of the teams and the increase in their autonomy usually facilitates communication, it is very important to establish the appropriate communication channels very clearly, and to be very aware of possible human problems (incompatibilities, varying attitudes to work and to cooperate, personal conflicts, *etc.*) in order to prevent them (*e.g.,* at the team-forming stage) or to face and cope with them quickly when they arise. Evidently, this is one of the main duties of the management – both the *formal* heads (hierarchical structure) and the *operational* heads (team leaders).

Industrial design is generally admitted to be the most important phase in the product life (birth). Every aspect of the future performance of the product is fully committed in this phase. Decisions in the early phases are quite important and have to be taken carefully since future consequences of wrong ones are very difficult to foresee unless a comprehensive and coherent design process had been set up.

Eco-decisions and environmental impact have for sure to be integrated into this bunch of vital decisions. Within this approach, the "Green Book on Integrated Product Policy" (IPP)[3] (European Commission 2001) should soon become the basic *Bible* for industrial product designers. This issue will be further discussed in future trends (see Chap. 6).

The objective of any design process should be to concentrate changes in the preliminary design stages in such a way that the product is mature by the time it reaches the manufacturing stage – in which changes bring heavy costs – and reaches the market in virtually "*untouchable*" form. Furthermore, any change when the product is already on the market is regarded as dramatic on account of brand image repercussions.

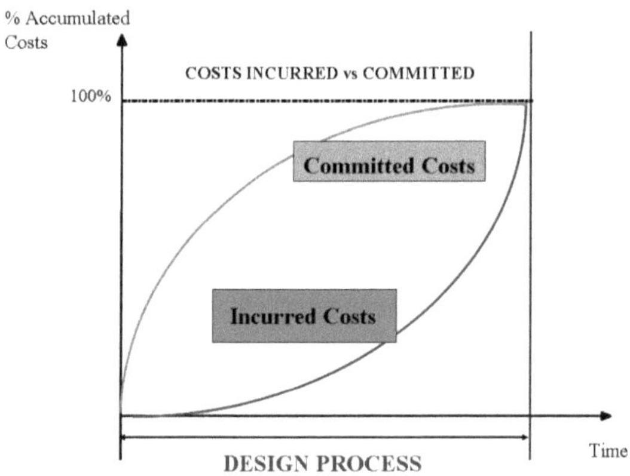

Fig. 3.2 Design costs

As can be seen in Figure 3.2, real costs incurred in the early stages of the design and development process are quite modest but the committed costs – costs that depend on decisions taken during the process and that "mark" the life of the product – actually "commit" the future cost of the product/process throughout its service life.

According to the integrated product policy (IPP) approach, it is evident that "life cycle costs" (not very much cared about until now) are soon becoming the most relevant cost item.

The new model to be described from now on covers all mentioned issues and provides a comprehensive framework for a consistent new product development and design facing the challenges of the twenty-first century.

---

[3] http://ec.europa.eu/environment/ipp/home.htm

Quality Function Deployment (QFD) as guiding thread for the model. QFD introduced in Chap. 1 (see Sect. 1.4.3), has to be used as the guiding thread for the whole process in the new model. QFD technique is extensively discussed in Sect. 3.3. The macro-flow (Fig. 3.11) and the horizontal deployment lines (Fig. 3.12) constitute the basis on which the new product has to be build by means of the described new model.

3.2.2.2 Stage 1. Company's Check-up

The purpose of the company's check-up is to perform an overall analysis of the "state of the health" of the company. Figure 3.3 shows the various techniques and tools to be used in the process. It is done at two levels:

- *Strategic level.* Strategic check-up
- *Technological level.* Technological check-up

The strategic check-up analyzes the strategic positioning of the company and its products as regards their present and future market. The approach should cover the global analysis of company strategy and its entire product portfolio.

In carrying out a strategic analysis of a company, the use of the Euro-Bunt[4] approach is a very good option. The Euro-Bunt (1995a, b) approach for strategic consultancy provides first, for a strategic analysis of the company on the basis of significant information on its internal situation and on the context in which it works; second, the approach proposes the implementation of action plans aimed at fostering the development of the company by introducing innovations of a strategic nature.

The strategic check-up should cover most of the company's areas, from the overall business idea down to the physical product and manufacturing process.

It supports strategic decision making on the range of: expanding the business concept to the "extended product" idea, *i.e.,*: jumping from the production of radio sets to the whole home entertainment business (Bang and Olufsen[5]); changes in production mix, changing activity to a new area based on the exploitation of the core competences, make or buy analysis, outsourcing, *etc.* coming in the ultimate case even at the decision of firm closure.

At the second level, the "technological check-up" will cover the whole life cycle of the product from conception to disposal. The technological check-up cares

[4] Euro-Bunt: this consultancy approach was developed in the first half of the 1990s by a consortium of a dozen European bodies taking part in the European Commission's Innovation program. As part of this project, a strategic management consultancy manual was published explaining the Euro-Bunt concept in detail: "The European Handbook of Management Consultancy. Strategic Innovation: An European Approach to Management Consultancy." One of the authors (M. Sorli) took part later in a project fostered by the Spanish Department of Industry to disseminate this approach on a Spanish national level.

[5] http://www.bang-olufsen.com

about the "operative aspects" related to product/processes. It produces accurate information upon a number of characteristics of the various product families in the company's catalogue: priority markets (portfolio), situation in life cycle, expected evolution in the short and medium term, profitability and its evolution, critical success factors, *etc.* The same goes for the processes: critical points, bottlenecks, waste identification, opportunities for improvement, technological trends, *etc.*

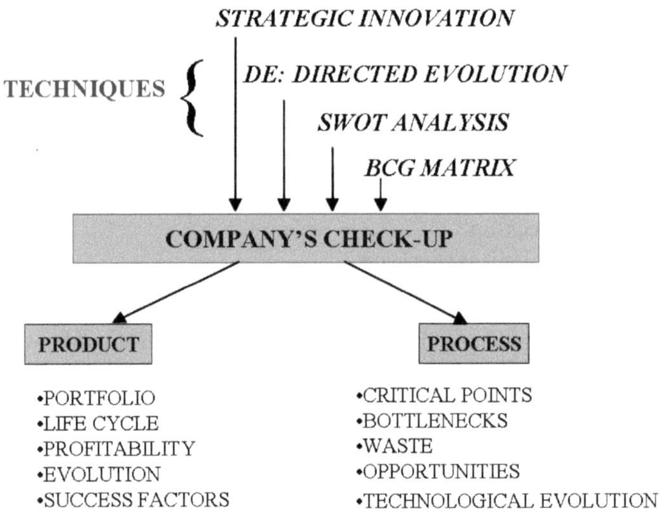

Fig. 3.3 Strategic check-up

3.2.2.3 Stage 2. Defining the Specifications

In this stage, the characteristics of the new desired product and the way it is to be manufactured are mapped out.

The key tool at this stage (see Fig. 3.4) is QFD which helps us to interpret and organize the various groups of requirements, the output being the set of product specifications. The inputs for the QFD process come from three main sources:

- The customer (the most important source)
- Internal needs or policies of the company (or EE)
- The situation of the competition

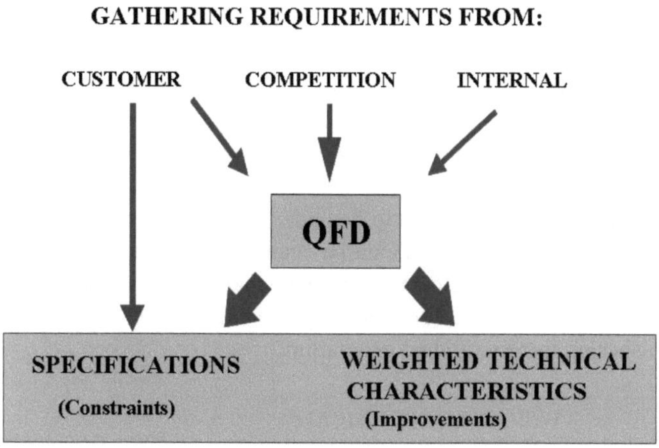

Fig. 3.4 Defining the specifications

According to the model known as the "Kano model" (Kano 1984; Sorli and Ruiz 1994), the specifications can be classified in three big groups:

1. Those termed "basic" or "restrictions" (left side of the block in the bottom of Fig. 3.4) are mainly directly imposed by the customer who sets down a number of minimum requirements in his tender specifications in the form of "constraints." These must be satisfied otherwise the purchase will not take place. For those products in the market place, "basic" characteristics are those that the competing products have already achieved in a highly reliable way being the failure, in consequence, quite inadmissible to the customers.
2. The "would-be improved" specifications (right side of the figure) are those that enable us to identify the main points which, if improved, will bring a significant increase in satisfaction to the customers in the market.
3. "Over-exciting" characteristics are those unexpected by the customer, which will naturally increase the satisfaction level exponentially. Not reflected in the box since they do not need to be weighted, nor improved, "being there" is just enough.

3.2.2.4 Stage 3. Conceptual Design

Conceptual design means designing a product in general terms, covering aspects such as the architecture and modularity of the system, its main and secondary functions, volumes, interfaces with other elements around it, *etc.* One or more possible conceptual solutions should be analyzed and selected in this conceptual design phase for subsequent validation.

However, it is highly advisable to define as little as possible at this stage as regards technological and functional solutions, materials to be used, processes, *etc.*, since if such aspects are fixed too tightly at this very early stage the range of possibilities will be cut down and creativity blocked, the result being routine, repetitive solutions – in a word, bad solutions.

As Fig. 3.5 shows, starting from the set of specifications, two distinct paths can be traced for reaching the conceptual design of the product: redesign (product improvement) or new design.

As previously mentioned, *new design* has to be understood as launching a radically new product,[6] so new that it may even entail a change in the line of business – making use of the organization's core competencies to change the product while staying with the same technology, or introducing a sizeable number of changes in the current product, ushering in new product generations.

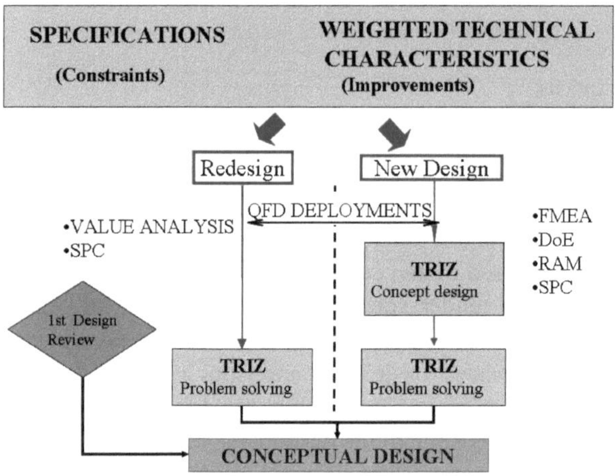

Fig. 3.5 Conceptual design

The distinction between *redesign* and *new design* is difficult to pin down generically, though designers clearly make the distinction for a specific product. In the automotive world, a fairly clear distinction is made between "restyling" involving a number of "cosmetic" changes made to the current model without changing its overall conception, and keeping the same name for the model, and "new model" usually creating a new name for it. In general, it can be said that the level of changes introduced is what makes the difference. As said before, the real point is customer perception: *if the customer feels that the product is really new, then it is new.*

Redesign (product improvement) can be viewed as developing a "new product" based on the current one through introducing a number of relatively small im-

[6] Though coming up with a "radically new" product is rather rare in our time, – now that everything seems to have been invented ("nothing new under the sun") – when a new release of any product features so many differences as compared with its current version, calling it a "new product" may be reasonable.

provements whereas "*new design*" implies developing and designing a significantly new product with a clear and tangible difference from the old one.

3.2.2.5 Stage 4. Detail Design

A variable and interactive set of activities runs between conceptual design and detail design, as shown in Fig. 3.6. They can be broken down into:

- *Simulation.* The use of tools, essentially computer tools, to "fill in" the conceptual design with technology and constructive solutions. Simulation and computation are used to validate the conceptual solutions, the most promising one being chosen.
- *Improving the solution.* Using conceptual tools to improve and optimize the chosen solution as far as possible.
- *Prototyping.* Developing real or virtual prototypes for evaluating the product performance in the real world as realistically as possible: structural representation, fatigue tests and trials, strength, look and feel, interferences with the surrounding context, ergonomics, *etc.*

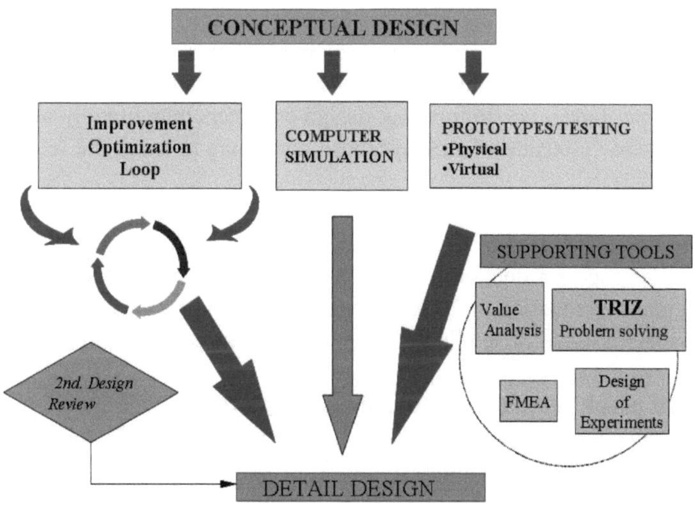

Fig. 3.6 Detail design

This process may go through all the interactions deemed necessary until a suitable solution is reached within the time limits set by the overall framework of the development process.

Innovation is very important in this phase supporting the development of new solutions to "fill in" the *concept design* building the *detail design.* Tools for innovation like TRIZ (see Sect. 1.4.4) are very useful in this phase.

3.2.3 Information and Communication Technologies (ICT)

It is obvious that only recent rapid evolution of the technologies of information and communication (ICT) has actually allowed this new working paradigm. As previously mentioned, Chaps. 4 and 5 deal in much more detail with ICT technologies so just a few comments will be advanced here.

Big manufacturers (*i.e.,* from the automotive sector) have for some time been launching many experiments on "virtual product development" by means of new Web based technologies in connecting remote locations in cooperative work on the same project.

Some aspects of the ICT tools have been previously mentioned, such as:

- *The use of systems* of the kind known as *shareware* and *groupware*; working on a common database, these facilitate a number of aspects such as sharing information, working simultaneously with the same data, off-site working, *etc.*
- *The existence of interfaces* for successful communication – interfaces that are not limited only to computer tools but rather are based to a large extent, and more importantly, on human relations.

Another important benefit of ICT for multinational companies, based not only in different countries but even different continents, comes from the possibility of working on a 24h shift. Passing the work packages over the *World Wide Web* (www) and using common workspaces and databases allows remote teams to jump from continent to continent, linking the project on a continuous mode, from USA to India, Europe, Japan, *etc.* In this way design engineers may literally work round the clock. These "e-business" strategies also contribute to lead-time reduction.

Some examples of this new ways of working can be collected from Honda claiming that the new 2001 Honda Civic model, the first model to use the new strategy, achieved a process reduction of 15%. Also Daimler-Chrysler, on a supply-chain pilot, claims a reduction of 92%, coming down from previous 14 days to just one, on the specific issue of sending production program information to suppliers.[7]

ICT give design people a sound base to build on. Nevertheless, one should not forget that tools of this kind are *just tools.* They need people working on them using methodologies and conceptual tools of the kind discussed throughout this book.

[7] Automotive Engineering International Magazine (2001). http://www.sae.org/mags/aei/

3.3 The 3 Cs Process: Customer Driven, Concurrent, Collaborative

The *new model* for product/process development processes has to take into account the 3 Cs approach: *customer driven, concurrent, and collaborative.* The three aspects are developed and analyzed in this section.

3.3.1 Customer Driven

Understanding that the design process has to be aiming to meet the customers' requirements and desires has taken a long time for most of the manufacturers and even now is not widely accepted.

Organizations today should be concerned with giving customers value. Value is the relationship between benefits (both tangible and intangible) the customer receives and the price he/she pays. When products are designed around customer needs, value and sales go up. When start-up problems are reduced and cycle times shortened, costs go down. Greater sales and lower costs mean greater profits.

Since 1966, the world's leading corporations have been improving new products and services based on what their customers value. These companies include big names in almost all kinds of products and brands all over the world such as 3M, Apple Computers, AT&T, Bridgestone, Chrysler, Ford, GE, General Motors, Hewlett-Packard, IBM, Intel, Kawasaki, Komatsu, Kubota, Matsushita, Mitsubishi, Motorola, NASA, Volvo, and hundreds of others in nearly every product, service, and even software category. One of the keys for their success is the use of QFD: Quality Function Deployment depicted in Sect. 1.4.3.

In the mid-1980s the slogan "quality is free" (Taguchi and Chowdhury 1999) was widely used. By focusing on customer needs, firms found that they could enhance their revenue by improving the quality of their products and services, understanding that the quality concept is related to the ability to fulfill market needs. Many firms studied their processes and found that they could improve quality by better tuning to the customer and at the same time reduce costs. "Getting it right the first time" reduced rework. Using reliable components reduced waste.

Quality function deployment (QFD*)* matches design decisions to customer needs. Quality was, in fact, free because firms were able to make their products and processes more efficient. Those firms that improved quality survived; those that did not either perished or were swallowed up by their more fit competitors.

As previously mentioned, QFD was born at Japan in the end of the 1960s as a mean to break the communication barriers among the different specialists in the Kobe shipyards of Mitsubishi Heavy Industries.[8] Akao, Mizuno y Furukawa

[8] Actually this experience was not exactly QFD but just constructing a matrix which is the basic tool for QFD. It can be said that it was kind of the "pre-history" of it.

(Akao 1990) helped the shipyard staff to develop a matrix in which to cross customer's requirements with quality characteristics. Shortly after, Dr. Akao founded and became first President of the "Committee for QFD Research" within the widely know and prestigious Japanese Association for Quality (JSQC).

Biased by its origin and Dr. Akao's background, early QFD models focused on the quality control on production systems but it was soon realized that the big potential for the tool was to be used starting early at the design phases, so enabling a real "*costumer driven design.*"

The unique distinctive characteristic of QFD is its focus in satisfying customers' needs looking even to exceed expectations by offering unexpected exciting performances in the product. The Japanese have issued the word *gemba* as the point where the product has value for the external world, where it is being used and experienced by real people. So the immediate consequence of this theory is to promote technicians and engineers to go to the *gemba* and actually understand how the product they have designed and produced is interacting with the users. As an old Indo-American saying states: "You will never know your enemy unless you walk for a mile in his moccasins" but in this case the customer is not the enemy but has to be the king to be served.

The QFD process uses matrixes as tools to cross different types of information and facilitate the people involved to deal with non-numerical concepts such as desires, conceptual requirements, and other customer's feelings.

A-1 matrix, following Bob King's terminology from his first English written book on QFD "Better Designs in Half the Time" (King 1989) or the *house of quality* according to a widely disseminated article "Quality by Design" (Clausing and Simpson 1990) is the most commonly used first matrix in every known QFD experience. A-1 matrix is a tool to collect, classify, and cross information. It plays a twofold role as a good information repository in which all team members may have the same set of information as well as a common understanding of it, and second, it also works as a huge Pareto[9] system in which the main issues are sorted according to their relevance to the customer.

Figure 3.7 shows the house aspect of the matrix (rationale for the name of the house of quality) and the content of each of the blocks can be seen in the figure.

Its functioning, in a few words, is to input from the left lateral side a previously filtered list of customer requirements (*what*) and cross it by the vertical entry of the technical engineering parameters (*how*) enabling the team members to understand the impact of each engineering characteristic in the customer requirements.

---

[9] Pareto is the name of a popular histogram (Pareto chart), coming from a nineteenth century Italian sociologist, economist, and philosopher: Vilfredo Pareto. The *Pareto chart* sorts any incidence (effect) by the accumulated number of times it repeats itself. It derives from Pareto's law of 80/20 stating that roughly 80% of the effects are caused by only 20% of root factors.

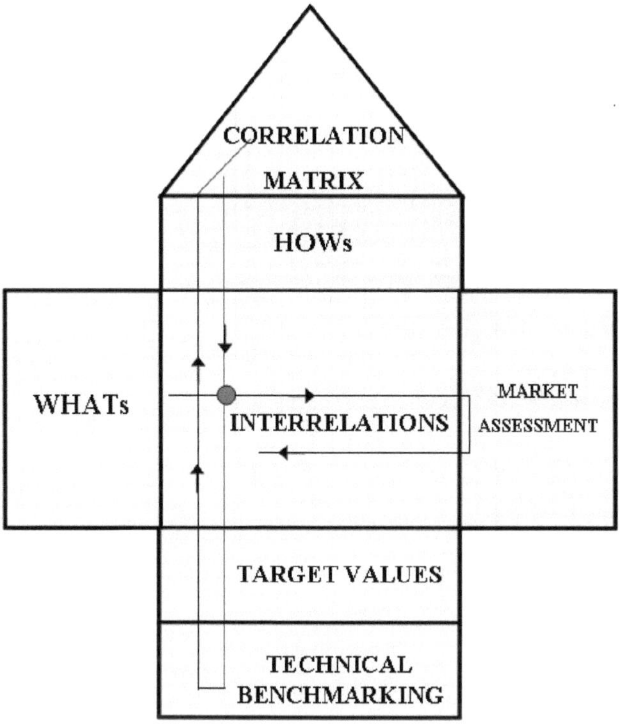

Fig. 3.7 QFD A-1 matrix

On the horizontal branch, the "*whats*" are sorted according to their weight for customer satisfaction and the comparative assessment with similar competitive products (right box). This classification is then transferred to the technical parameters in the bottom box where a technical comparison (benchmarking) is also performed. Based on that, the priorities are set and target values assigned to each of the selected parameters.

Finally, the top of the matrix (house's roof) helps the designers to identify positive synergies (to benefit from) or possible conflicts (to prevent) between parameters.

Figure 3.8 shows the general aspect of an experience of this kind (Sorli and Ruiz 1994). The matrix is just an academic exercise and represents a QFD analysis for a wrist watch. The text is in Spanish and the matrix has been built using QFDcapture™ [10] software.

[10] QFDcapture™ is a trademark of International TechneGroup Incorporated, 5303 Dupont Circle, Milford, OH 45150, U.S.A. http://www.iti-global.com

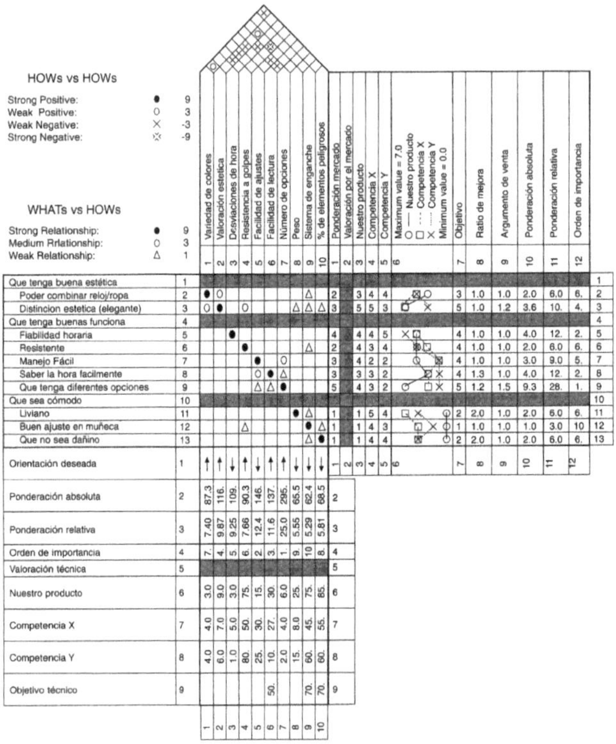

Fig. 3.8 QFD A-1 Matrix for a wrist watch

This A-1 matrix is very valuable to enable technical people to understand the customer perspective; both approaches, as shown in Fig. 3.9, are quite different.

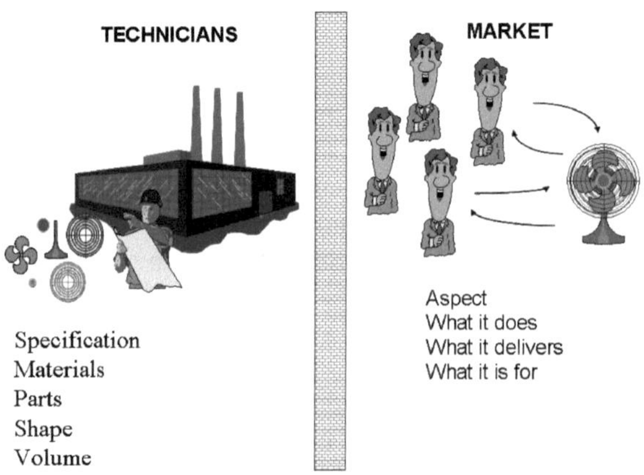

Fig. 3.9 Different perspectives from technicians and market

However this matrix should be completed by incorporating several other requirements coming from internal sources (maintenance, production, logistics, *etc.*) or external to an upper level to the market (legal requirements, societal need, *etc.*) in the *enhanced matrix* (Fig. 3.10).

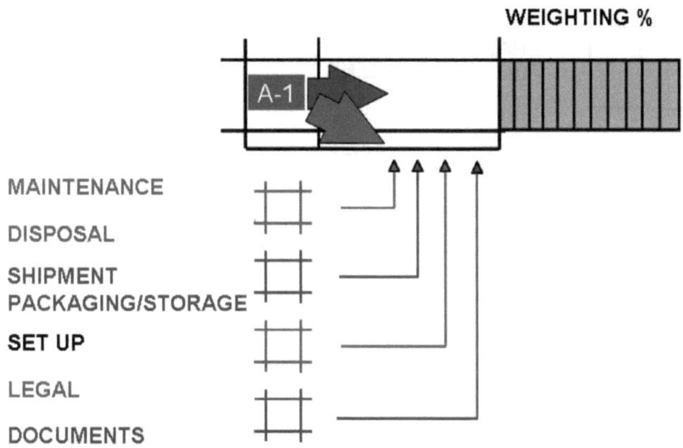

Fig. 3.10 Enhanced matrix

In that way, an overview of the product/process requirements is compiled in a single "database" that is intended to cover the whole life cycle of the product. As has been presented in the stage 2 of the new model (Sect. 3.2.2.3), output from matrix A-1 and enhanced matrix will constitute the set of specifications of the product giving an answer to the question: what and which level of performances will the product deliver?

Starting from the A-1 Matrix (starting point for all deployments) QFD macro flow (Fig. 3.11) collects and analyzes customer requirements and keep them alive along the whole process to the production phase. From the above discussed first matrix (A-1) the QFD macro flow drags the selected parameters or "performance indicators" (priorities) down the flow from the product (phase I) to its systems/parts breakdown (phase II), next to the process planning (phase III) to finalize on the production control parameters (phase IV). For sure, each phase may consists of just one matrix or be split into several depending on the depth of the analysis and some other factors.

Macro flow vertically transfers customer requirements to the bottom line in the factory, creating a consistent link between any decision on the shop floor and related customer needs.

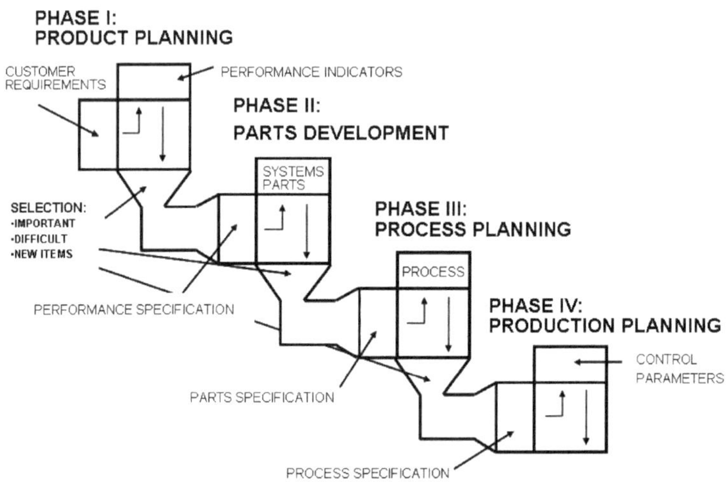

Fig. 3.11 QFD macro-flow

This vertical QFD macro-flow can be complemented by the "horizontal four deployments" as shown in Fig. 3.12 where within a given time frame, the four main performance characteristics of the product are fixed. QFD deployments work on a horizontal scope covering the whole product definition, including environmental issues, on the four main aspects of:

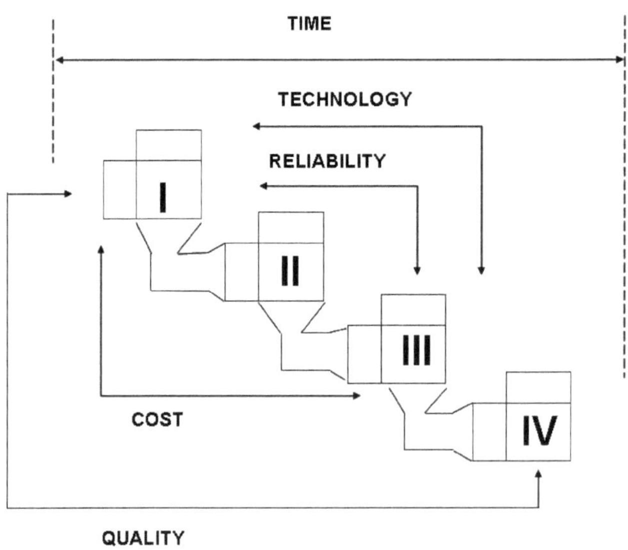

Fig. 3.12 Horizontal QFD deployments

- *Quality*. What the customer expects and how will the manufacturer fulfill it? Environmental issues have to be considered as one of the most important aspects of quality: "*eco-quality.*"
- *Reliability*. How is the product going to perform with time and what is its life expectancy? Sure enough, environmental features have to be kept along with time in the pre-defined range.
- *Cost*. Is all this going to be achieved within the cost objectives? Eco-design considers "cost" of the product/process as the overall cost for humanity in terms of resources consumption and impact on the environment.
- *Technology*. Available technology is allowing us to do everything as planned, or should we try something new? Different technologies have different impacts on environmental aspects.

Answers to these questions have to be discussed by the designer's team in an iterative mode within time constraints limiting the process. If any of the answers is negative, new assumptions have to be made, some objectives have to be downsized, or...innovative solutions have to be developed and applied. Iterative analysis of the four objectives may eventually come back to revision of the target values established in matrix A-1.

Integration of QFD and Value Analysis (VA). Conjoint utilization of QFD, and VA supported by the functional analysis – *i.e.;* through "function analysis system tree" (FAST) – in the product design process provides a powerful dynamics assuring a comprehensive and consistent product in which "everything" has been taken into account (Sorli and Mañà 1997).

This integration can be developed in three different ways as follows:

1. *Product specification.* QFD matrix A-1 (see Fig. 3.13) is built in order to analyzing customer requirements (whats) and identifying performance measurements (hows) to fulfill these requirements. This A-1 matrix should also be completed using instead the extended matrix gathering requirements not made explicit by the customer/market: internal requirements, manufacturing needs, legal and company requirements, *etc.*

On the other hand, *function analysis* (see Sect. 1.4.6) provides the upper entry to the F-1 matrix being the lateral entry with the same input that in the previous matrix (A-1). This *function/what* crossing allows the team to complete the listing both of *whats* and *functions* so detecting possible missing requirements or functions.

The following matrix (F-2) has the aim of issuing all the *performance measures* (hows) related with the list of *functions*. Every function gets a measure item that will allow us to monitor the performance of this specific function. In function analysis methodology, "hows" (measurements) are also known as "criteria."

Both from A-1 and F-2, a complete listing of all performance measures (hows) with assigned objectives is obtained. The collection of all these "hows" constitutes the "*product specification.*"

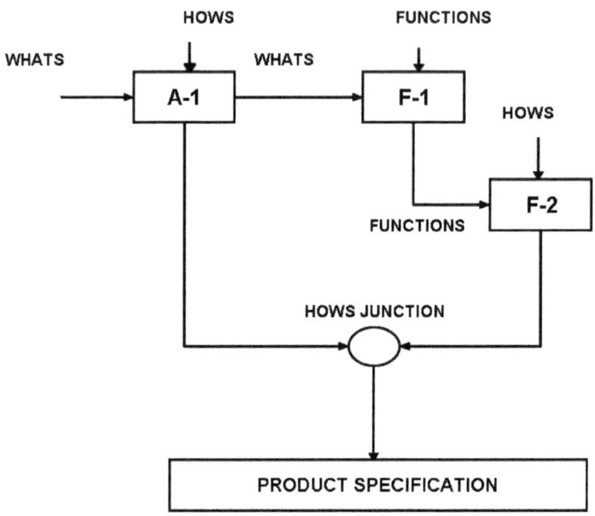

Fig. 3.13 Defining product specification through QFD process

2. *Cost assignment to functions.* In this second approach (see Fig. 3.14), the QFD mechanics transfers the customer weight from the requirements to the performance measures. (matrix A-1).

In that way, the "value" the product gives to the customer/market has been translated to engineering performance items. That's to say which items have the biggest impact in customer appreciation.

The next step is to transfer this customer value to the functions. Considering 100% as the total value of the product for the customer, each of the functions has to cover its own portion from the total. But the important issue is that in order to be effective, functions have to assume a percentage of the total forecasted cost parallel to their contribution to the value.

Cost assignment to functions may be reached using any of the options showed in the figure:

- *Option A.* Directly from matrix F-1 *customer requirements (whats)*
- *Option B.* From matrix f-2 *performance measurements (hows)*[11]

[11] f-2 is the reverse matrix to the F-2 of the previous figure.

The value from both alternatives has to be very similar provided the analysis has been done correctly. The idea of this double-crossing (if feasible) is to help the analyst to detect incoherence and deviations in the study.

From there on, there is a link with the value management (value analysis – see Sect. 1.4.6) methodology that will enable designers to identify critical points and subjects for improvement.

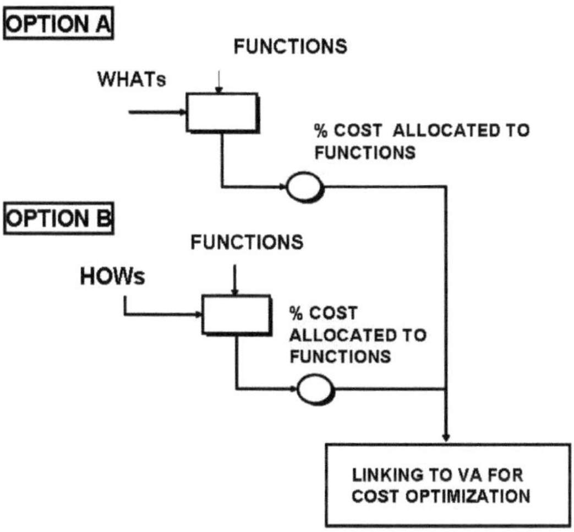

Fig. 3.14 Costs assignment by functions

3. *Cost assignment per item.* Finally, the third alternative is to distribute the cost per elements of the product (Fig. 3.15). Having established an objective for the overall cost of the product (design for cost objective), this cost is distributed to elements at different levels of the product depending on its complexity, or the level of detail to be dealt with: systems, mechanisms, sub-assemblies, parts.

The total cost of the product is given by the sum of cost of raw materials plus manufacturing assembling of every part.

Obviously, some of the components or parts will just be *raw materials* (no manufacturing cost) and the final assembly will have only manufacturing cost.

Both QFD and value management propose the basic idea of finding the balance between the cost of every component (including both factors) and the participation of the same component in adding value to the customer. This comparison may be achieved using percentage values considering that the global product has 100% cost (no matter the real figure) and provides 100% satisfaction to the customer covering a list of requirements (*whats*) that have their percentage of participation on 100% satisfaction.

The ideal situation will then be that the “value index for component” (I_{vi}) be equal to the unit:

$$I_{vi} = \text{Contribution of component “i” to value /Cost of component “i”} \quad (3.1)$$

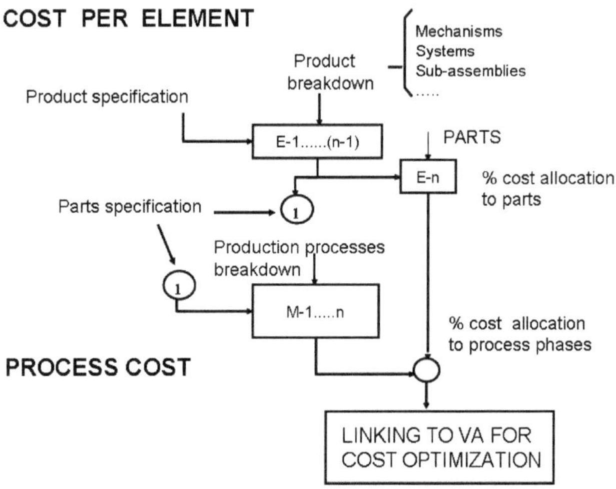

Fig. 3.15 Costs assignment by product element and process phase

Following the flow in Fig. 3.15, the percent value contribution of every element may be calculated up to the desired detail level building the necessary matrices. (E-1,....E-n).

In parallel, contribution from every operation from the process (manufacturing or assembly) can be evaluated using process matrices. (P-1,.....P-n).

From this process the contribution of every element and its manufacturing process has to be matched to the customer value. Based either on current or objective costs the balance “contribution to the value/cost” is calculated.

As has been said, the ideal situation is when the value index for element (Ivi) is equal to the unit. Ivi having values very separated from the unit can be considered as critical items to improve. For these elements (may be assemblies, single parts or either process phases) value analysis methodologies are again applicable.

Output. At the end of this process a specific cost objective is assigned to every item in relation to its contribution to the value of the whole product for the customer.

Components/elements with a poor balance are clear subjects for improvement both by increasing contribution to the customer’s appreciation or through reducing manufacturing costs.

Several new concepts for these required modifications have to be issued and evaluated against established criteria based, as usual, on customer’s rating.

3.3.2 Concurrent Engineering

Concurrent engineering is the new trend superseding the traditional sequential system presented in Chap. 1 (see Sect. 1.2).

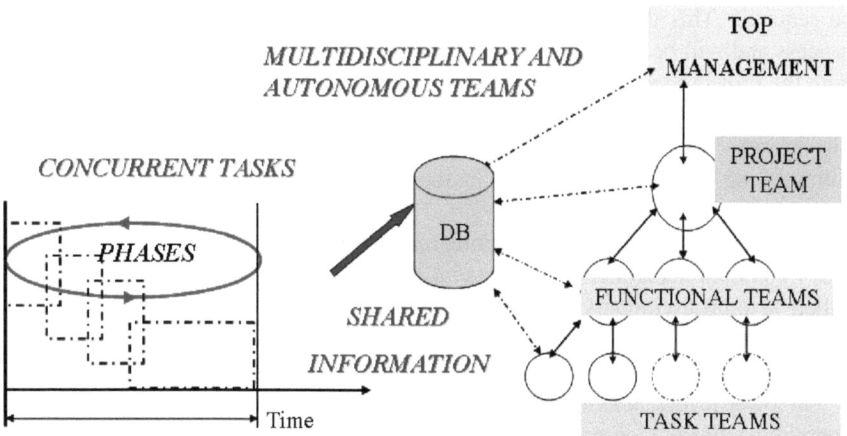

Fig. 3.16 Concurrent engineering

As can be seen in the left side of Fig. 3.16, working concurrently has several implications. The most important without doubt is the organizational change described below. Three main blocks can be differentiated in this concurrent engineering approach:

1 Time schedule. (Gantt chart on the left of Fig. 3.16). On one side project tasks are not only overlapped in time but they have to run in parallel to the highest possible extent in an interactive and integrated manner.

This approach is very much related to the new organizational set up but also requires a change of mental paradigm. Traditional perception of tasks that can't start until receiving results from a previous one must shift to consider that the overall process is the responsibility of the project team (to be described later) independent of the specific tasks each team is committed to.

Under this new approach, provided that the tasks relationship is quite understood and teams are working in an integrated manner, being conscious of the whole picture and making contributions to the totality of the process, all tasks should be considered as starting at once. This means that, even if no physical activities are undertaken, those responsible will follow up the process from the beginning and will perform any necessary preliminary preparatory activities.

2 Organizational structure. (Right side of Fig. 3.16). To achieve this paradigm, the organization must change from hierarchical to project oriented following the structure represented in Fig. 3.16 (right side). Furthermore, this structure is not only internal to the company but has to integrate representatives of the EE: *i.e.*, members of the product value chain.

A specific *Project Steering Team* with a high level of autonomy has to be created, having full responsibility and decision making capacity and reporting directly to top management from which only a selected number of decisions should be required. This decision points will be scheduled in the overall development process and will be performed through decision making meetings of management with the Project Steering Team.

In that way, one of the main causes of delays – time for decision making by the big bosses – would be eliminated provided that the time schedule is adequately followed up and decisions made in the right time with the right information needed.

The Project Steering Team will integrate representatives from different companies traditional departments with several expertise and knowledge. This is what is called "a multidisciplinary and multi-functional team."

This Project Steering Team will be stable throughout the project duration and will disappear or change for new projects or missions for its components.

On lower levels, *functional* and *task teams* will appear and disappear on a work completion basis. Composition and characteristics of these teams will depend greatly on the type and complexity of the products. These teams will care about specific functional assemblies of the final product (*functional teams*) or specific tasks (*task teams*). They will have the same characteristic of being autonomous, reporting to the steering team to which some specific decisions will be delegated in the same manner as described above regarding the relation between the Project Steering Team and the top management of the company.

Furthermore, members of the Project Steering Team will integrate second level teams, and members of these second level teams will form part of the lower level ones. In that way, knowledge, information and any other types of human formal or informal communication will flow smoothly among all teams, allowing a global knowledge of what is going on.

3 Information and communication technologies (ICT). This new depicted scenario would be very hard to achieve without support of advanced ICT means. It is clear that this new way of working implies new situations resulting from decentralized people working at the same time on the same pieces of information, and this needs to cope with aspects such as:

- A common repository where shared information should be updated on real time and put to disposal of many different actors on teams
- New means of control of the working schedules of the different teams, control of the whole process, engineering changes, communications, *etc.*

Access restrictions and confidentially aspects are other issues to be taken into consideration. On the policy of autonomous self-sufficient teams, restrictions should be minimal but in any case, due to the common work of agents from different actors in the value chain, it is an issue that has to be analyzed and very carefully dealt with. All these aspects will be discussed in detail in Chaps. 4 and 5.

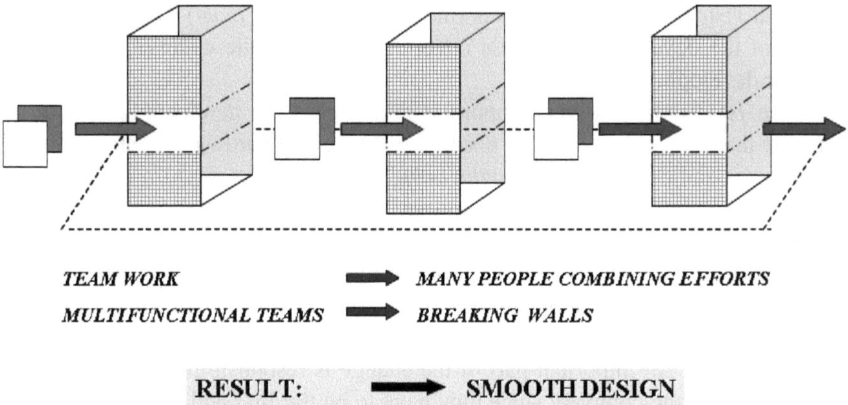

Fig. 3.17 New scenario by concurrent engineering

Linking with the previous discussion on industrial evolution (Sect. 1.1), concurrent engineering is intended to mitigate the problems of the modern situation looking for a return to the initial craft era as shown in Fig. 3.17:

- The issue of a high number of people involved in the process is managed by means of team working combining team's expertise, people's different views and knowledge
- "Design over the wall" is superseded by breaking barriers between functions and specializations through the multifunctional teams

Keys of concurrent engineering. Concurrent engineering new paradigm is mainly supported by some key issues that may be cited as:

- *Integration of efforts* by means of multifunctional teams
- *Management of information* based on a good project planning supported by ICT
- *Mastering the technology* which in the complex and turbulent current world may only be achieved within the EE frame and the multifunctional team concept
- *Fostering innovation* as the only effective mean of problem solving and continuous improvement
- *Common sense.* A correct functioning of teams allows a better exploitation of individual common sense which in standard hierarchical structure, on the contrary, faces more obstacles

- *Caring about detail.* Common work in teams, more relaxed working conditions and adequate time span allows devoting some time to minor details that, as mentioned before, may generate great positive or negative effects
- *Global vision.* Members of teams (mainly the *steering team*) have a global vision of the development process covering the whole life cycle of the product under development
- *Means provision.* One very important thing about the new paradigm is to be aware that new means are also needed: software and hardware solutions, methodologies and over all them: *human beings*
- *People.* Persons working on concurrent engineering environments cannot be the same[12] as used to work on traditional conditions. Persons released to work in a team of this kind have to be relieved of their former tasks/responsibilities and their performance indicators and motivation factors have to change accordingly. Team leaders must be real "leaders" and not necessarily official bosses
- *Management.* An important part of the change is responsibility of the management staff of the company since they have to be the first and more enthusiastic driver of the changes and also have to set up the employees' motivation and reward mechanisms that would achieve a smooth functioning of such a complex new organization

Extended Enterprise. A very relevant aspect of the concurrent engineering paradigm is the inclusion of the Extended Enterprise (EE) concept discussed in Sects. 2.3 and 3.2.1. This concept encompasses all agents participating in different phases of the product life cycle and production process.

Within the EE, one of the most important actors is the group of suppliers. Historical evolution of the supplying system has evolved from the old times in which the purchase department just bought raw material and other components to the more economic supplier to the idea of "comakership" in which the first tier suppliers assume the coordination of a cluster of second tier suppliers and owns the development of a whole assembly which has been developed in joint co-operation with the final product manufacturer (Merli 1994; Sorli and Gómez 1994) – see Fig. 3.18.

The main changes in this evolution process can be highlighted as:

- Individual supplier of one single part becomes responsible for joint development of a complex assembly to be integrated in the final product: first tier supplier
- From just providing a single part upon producer requirements, the first tier supplier develops the whole product fitting into the overall requirements of the buyer: original equipment manufacturer (OEM)

---

[12] Certainly staff cannot be changed overnight but the same people have at least to change their minds and adapt themselves to the new world.

- From a myriad of small providers of a wide range of products, the first tier supplier becomes the owner of the assembly and full responsibility of lower tier suppliers that will procure him parts and sub-assemblies. He is responsible to the whole process, logistics, and delivery of the parts following the just-in-time (JIT) approach

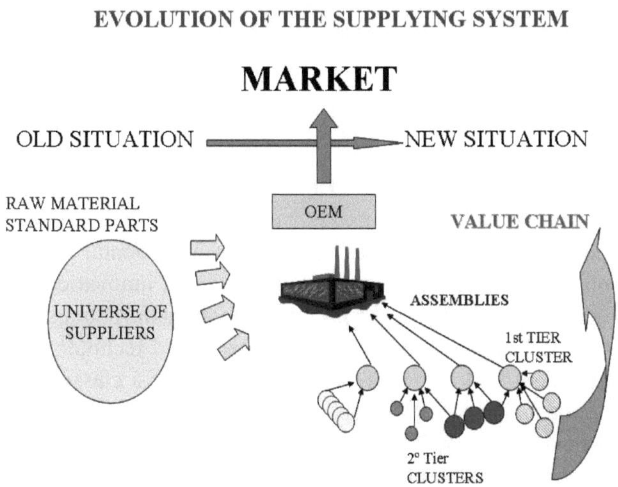

Fig. 3.18 Comakership

From the idea of the lowest cost suppliers of simple parts, whom through time were also required to fulfill other requirements on quality, delivery time, financial stability, reliability, *etc.*, first tier supplier becomes a "*comaker*" in the sense of long term business relationships based on a mutual trust and *win-win* approach.

3.3.3 Collaborative Working Environments

Collaborative product design and manufacturing among distributed teams through Web based tools and services is becoming more necessary as enterprises are distributing their activities throughout the world, in many cases working under the round the clock approach, that is to say, working 24 h a day, passing the work-log from one team in a darkened continent to another team in a different region of the world where the sun is now rising to start a new day.

The support for integrated teams' creation through an integrated and well tailored ICT approach can lead to crucial advances in the business area. Application of state-of-the-art ICT solutions is necessary to assure higher efficiency of the cooperation and integration processes.

The systems should support the integration, management, and reuse of knowledge *via* a common knowledge base, in a form of essential expertise, reachable anywhere, at any time.

Collaborative Working Environments (CWE) are an emerging issue that is becoming more and more relevant in the way companies in the value chain (EE) are working nowadays (Mendikoa *et al.* 2008). This point will be discussed in more detail further on (Chaps. 4 and 5).

3.4 Systemic Innovation

Modern companies have to meet the challenges of increasing product variants and service content of products required on the market by innovating their products and business processes in a new way. Instead of the classical incremental innovation approach, new forms of innovations are needed. The product innovation and business processes innovation in industry in the past were done by introducing advanced technology in products and new organizational forms and technology in certain processes based on identified demands. This corresponds to a classical incremental innovation approach. The revolutionary step in product/process innovation in industry is to "radically innovate" the whole innovation process itself and the whole industrial working environment, by focusing it upon the main actor in industry, the human actor, and by applying the emerging *systemic innovation* approach.

This section is dedicated to the *systemic innovation* paradigm as one of the key innovation approaches in industry in the twenty-first century and is mainly following the study provided by Maula, Keil and Salmenkaita (Maula *et al.* 2006). During the past decades of the twentieth century such innovations have become increasingly common including the Internet, the 3G[13] mobile telephony, Linux®[14], Java, Symbian, and many others.[15]

As indicated in previous chapters, several types of so-called non-incremental innovation have been identified that create particular challenges for the product/process innovation in industry:

- *Radical innovations* that change core technical concepts and their linkages (Tushman and Anderson 1986)
- *Architectural innovations* that change the linkages between core concepts (Henderson and Clark 1990)

---

[13] *3G* stands for the *third generation* of standards and technology in mobile telephony. It is based on the International Telecommunication Union (ITU) family of standards under the IMT-2000.

[14] Linux® is the registered trademark of Linus Torvalds in the U.S. and other countries.

[15] Open source software for mobile and Web based applications.

- *Disruptive innovations* that address new customer groups and focus on different performance characteristics (Christensen and Bower 1996).

These three groups of innovations have in common that they require organizational arrangements for innovation that differ from processes suitable to create or adapt to incremental innovations (Maula *et al.* 2006).

3.4.1 Definition

Systemic innovation is a new approach for understanding innovations occurring in an economy. This approach has emerged during the last decade of the twentieth century. It points to the fact that innovation processes are evolutionary, and does not therefore make use of the notion of optimality. It also stresses that firms do not normally innovate in isolation but in interaction with other organizations within the framework of specific institutional rules.

Systemic innovations are based on a combination of new technology and needs created, innovations in the process that require multiple companies to change their practice, typically enabling significant increases in overall productivity over the long term (Edquist 1999, 2001).

There is no unique, well established definition of systemic innovation. Conceptually such systemic innovations were introduced to the literature of innovation management as a category of innovations requiring specialized complementary assets for successful commercialization of the innovation in question (Teece 1986, 1996). Systemic innovation has been defined as an innovation whose "benefits can be realized only in conjunction with related, complementary innovations" (Chesbrough and Teece 1996). There are several key features of systemic innovation which will be examined in the following text. These features are:

- *Non-autonomous character of systemic innovation.* Autonomous innovations are defined as those that can be pursued by companies independently from other innovations or from other companies (or production stages). Innovation is considered systemic if it requires the coordination of change across more than one production stage (Langlois 1992) or requires coordination throughout the system in order to realize the gains from innovation (Teece 1996). Systemic innovations thus have to be distinguished from autonomous innovations that "can be pursued independently from other innovations" (Chesbrough and Teece 1996). Systemic innovation (*vs* autonomous innovation) requires significant adjustments of other parts of the business system they are embedded in (De Laat 1999; Teece 1986, 1996).
- *Coordinated and networked innovation.* Systemic innovations require significant coordination because they must be carried out with related or complementary innovations. Due to the fact that systemic innovation processes frequently span beyond the boundaries of the firm they often entail the coordination of dif-

ferent parts of the value network and entail open innovation organization models of innovation activities (Chesbrough 2003b).

- *Open Collaboration.* Systemic innovations cross stages of production or require adaptation throughout the entire "system." Because information sharing and coordinated adjustment are required for fast and efficient problem solving with systemic innovations, the limited administrative control and high powered incentive structures of markets may impair the adaptive and sequential changes that are needed. Instead, systemic innovations require institutions with low-powered incentives, where information can be freely shared without worry of expropriation and disputes between economic actors can be easily monitored and resolved in a timely and efficient fashion (Teece 1996).
- *Creation of needs.* Differently from classical innovation approaches, which always tend to response to the needs identified at the market and business processes (*e.g.*, problems identified), systemic innovation may also involve proactive creation of needs, *i.e.*, the companies may actively work to create needs in the market.

Systemic innovations typically enable significant increases in overall productivity over the long term. But these may create switching or start-up costs for some participants and reduce or eliminate the role of others. Examples of systemic innovations include virtual design and construction, supply chain integration, and in homebuilding, prefabricated subcomponent wall systems (Taylor *et al.* 2004).

In systemic innovation, companies need new tools for foresight and shaping to manage the business environment of the corporation over different time horizons. This increases the role of tools such as external venturing, research collaboration, and industry consortia.

Although systemic innovation has been practiced in industry for more than a decade (especially in the ICT industry), Maula, Keil and Salmenkaita (Maula *et al.* 2006) concluded that systemic innovations have been subject to very limited discussion in the literature. The systemic characteristics of innovations have been identified to impact selected business dimensions of innovative activity.

As indicated above, it should be noted that there are certain differences in definitions of types of innovation. For example, Henderson and Clark (1990) defined the concept of architectural innovation and investigated several seemingly straightforward innovations that resulted in significant consequences for the photolithographic alignment equipment industry. Their goal was to understand what characteristics of those innovations were unique. Henderson and Clark's research suggested that the linkage between the core concepts and the components in a product or process innovation were important factors in describing the landscape of types of innovations. Taylor, Levitt and Raymond (2004) describe practically the same type of innovation as systemic innovation.

3.4.2 Coordinated and Networked Innovation

As explained above, one of the key features of the systemic innovation is that it requires coordination and networking. In systemic innovation processes companies also need to network and coordinate with producers of complementary products (and processes) and in many cases even with direct competitors to ensure the viability of the innovation, rather than coordinating solely with suppliers and customers as is frequently the case in closed innovation models. While systemic innovation processes are widely practiced in industries such as telecommunications or information technology, the processes how incumbents and new entrants achieve this coordination and ultimately how they jointly create systemic innovation are not always well understood.

Although many studies have examined innovations that could be characterized as systemic (see Sect. 3.4.1), these studies have not systematically analyzed the coordination of proactive creation of the entire system of innovations both in products and processes. The studies have frequently taken as given the long and evolutionary development process of the complementing innovations, such as the development of the petrochemical industry to provide fuel for the combustion engines of automobiles or the development of the production and distribution of electricity to enable electric light to displace gas lamps (Utterback 1994).

However, complementing innovations are critical aspects of systemic innovations, and in the current business environment companies rarely have sufficient time to wait for the emergence of such essential complementary resources. The companies have to lead the development of systemic innovations proactively.

The distinction between individual autonomous or systemic innovation and the broader system has to be clearly identified. In "Open innovation in systemic innovation contexts" (Maula *et al.* 2006) it is stated:

> The characterizations tend to classify innovations as linked to either one firm or one product or technology category, forcing the analysis to extend to a more complex environment in terms of organization and dynamics of innovations. Literature that has directly focused on systemic innovations has largely focused on whether systemic innovation should be managed within a single firm by vertically and horizontally integrating complementary innovations or whether these innovations are better created through markets. Teece (1986) as well as Chesbrough and Teece (1996) have argued that systemic innovations should be typically managed in an integrated fashion to avoid the substantial difficulties in coordinating the innovation activities of multiple players in the market place.
>
> However, this view has been seriously challenged in some contexts *e.g.*, by De Laat (1999) with the argument that many contemporary systemic innovations are just too big and complex even for the largest integrated companies to manage alone. While integrating systemic innovation economizes on the cost of coordination and provides control benefits, it is frequently infeasible since even the largest firms lack the financial resources let alone technological and market capabilities to create the simultaneous complementary innovations necessary for successful systemic innovation. Empirical evidence supporting the integration argument is rather inconclusive, limited (Teece 1996) and at least partly

> contradicting. Observations concerning for example the telecommunications and Internet technology industries since mid-1990s present several examples of highly systemic technologies pursued through various types of collaborative efforts by a number of firms (Kano 2000; Keil 2002).
>
> Maula *et al.*

According to Taylor (2005) the degree to which outcomes are impacted by a systemic innovation is determined by the values for the four mediating constructs:

- Strong relational stability
- Network-level interests
- Fluid boundary strength
- Agent for network-level change. (The existence of such an agent will mitigate difficulties of mutual adjusting to a systemic innovation)

Conversely, weak relational stability, firm-level interests, rigid boundary strength, and the absence of an agent for network-level change will exacerbate difficulties of mutual adjustment and slow the diffusion rate for a systemic innovation.

3.4.3 Collaborative Aspects of Systemic Innovation

Systemic innovation makes companies increasingly dependent on others. Because in complex systemic innovations, vertical integration is rarely an option, innovation processes become increasingly collaborative processes (Maula *et al.* 2006).

Innovating companies are dependent on complementary innovators. The move from internal innovation processes to collaborative open processes forces companies to take a wider perspective to resource allocation and to adopt new governance modes to carry out activities related to the creation of systemic innovations.

It has to be understood that each collaborative innovation process *per se* is *not systemic innovation*, but it is very likely that each systemic innovation requires collaborative innovation processes. Even classical incremental innovation could be, and often is, carried out collaboratively: it may involve different departments in a company and/or suppliers in solving a specific problem and innovation of a product or a process. However, it could be carried out within a single department (*e.g.*, product design department may refine a certain component).

Systemic innovation essentially requires collaborative work. As indicated above, systemic innovation requires:

- Collaborative work within a company – involvement of many departments within a single company
- Collaborative work with a network/supply chains, even with competitors in order to coordinate the innovation process and synchronize complemented innovations needed for successful systemic innovation

Obviously as the systemic innovation requires involvement of many actors, there is a need for much more powerful and efficient tools to support collaborative work than in classical incremental innovation. As indicated in Sect. 3.4.2, the companies need tools to lead systemic innovation proactively by effectively coordinating collaborative work of many involved actors. This is the reason why systemic innovation is often confused with collaborative innovation processes. In Chap. 4 the ICT-based tools to support collaborative work are discussed, and their application for collaborative product/process innovation is elaborated in detail in Chap. 5. Although the solutions to be discussed are valid for any collaborative innovation process (even collaborative incremental innovation), their full benefits are to be expected within systemic innovation processes.

3.4.4 Resources for Systemic Innovation

Resource allocation is critical for systemic innovation processes. According to Maula *et al.* (2006), resource allocation in innovation processes has received attention in the strategy, technology, and innovation literature. Starting from the general resource allocation model developed by Bower (1970), this literature has evolved through a cumulative body of research over a 30-year period (Bower 1970; Burgelman 1983a, b; Gilbert 2002; Christensen and Bower 1996; Noda and Bower 1996).

The Bower-Burgelman process model, named after Bower (1970) and Burgelman (1983a), views resource allocation as part of a larger strategic management process conceptualized to consist of multiple, simultaneous, interlocking, and sequential activities that take place on the front-line, middle and top management level of the organization. In "Open innovation in systemic innovation contexts" (Maula *et al.* 2006) is indicated:

> Within the system innovation process the organization has to take strategic decisions and in particular resource allocation decisions. The central feature of the Bower-Burgelman model is that strategic initiatives emerge predominantly from the activities of front-line managers and then compete for resources and top management attention. In later work, Burgelman (1994) shows how the process of the emergence and selection of initiatives can be understood in an EE or a networked enterprise perspective as a variation, selection and retention framework. Noda and Bower (1996) further extended the model by showing how iterative processes of resource allocation lead to escalation and de-escalation of strategic commitments.
>
> Maula *et al.*

However, the Bower-Burgelman model appears to be inadequate for some innovation types. Different perspectives are needed to address resource allocation for different types of innovations. For instance, resource allocation along the lines of the Bower-Burgelman model fails for disruptive innovations (Christensen and Bower 1996). Disruptive innovation requires changing/adaptation of traditional metrics of product/process performances and business models. Disruptive innovations differ from incremental innovation in that they lower product performance along traditional metrics, but find an untapped need with a new set of applications, and find a broader set of new and initially different customers, who value these attributes and applications, and creates a significant change in the underlying business model of the firm, often lowering gross margins or changing the basic drivers of firm profitability (Christensen 1997; Christensen and Bower 1996; Gilbert 2002).

Resource allocation in line with the Bower-Burgelman model frequently fails to support such innovations since they do not fit the financial and operating criteria required to sustain the core business. The problem for incumbent firms is that, when disruptive proposals are considered, analysis based on established performance criteria reveals the new opportunity as inferior when compared with other potential opportunities that sustain the existing business. Gilbert (2002) complements this perspective by arguing that organizations frequently need to employ different and changing cognitive frames for disruptive innovations. Yet others have suggested separating the development of disruptive innovations in new venture divisions with separate resource allocation processes to enable them to survive in organizations (Tushman and O'Reilly 1997).

The research on disruptive innovation suggests that different innovation types might require differing resource allocation logics. For corporations to be able to develop radical or disruptive innovations, the prescription has been to establish separate new venture divisions to insulate the immature disruptive ventures from the pressures of the core businesses and thereby to create space for the long term development of more explorative ventures that are critical for the long term competitiveness of the firm.

However, Maula, Keil and Salmenkaita (Maula *et al.* 2006) argued that these prior resource allocation models optimized for allocating the internal resources of the corporation are insufficient for systemic innovation. For corporations to be able to create systemic innovations, yet further development is needed in the resource allocation processes. In prior research into the resource allocation processes, the resource allocation deals with allocation of internal resources, *i.e.,* employees, machinery, financial resources of the focal company. However, for innovation processes that require multiple simultaneous innovations in independent companies, such a perspective is too narrow.

In systemic innovations, partners and external developer communities make up a significant resource pool working on developing different components of the systemic innovation (Von Hippel and Von Krogh 2003). These developer communities and external partners are critical to the success of the innovation in question but are not under direct control of the focal corporation. Attracting and retaining the commitment of these external resources is the key to building systemic innovations proactively.

For creating disruptive innovations, the recommendation is to establish separate new venture divisions. However, optimizing the allocation of internal resources in a corporation may lead to sub-optimization when viewed from the perspective of creation of systemic innovations. In systemic innovation, resource allocation is not only about one's own resources. For successful proactive management of systemic innovations, resource allocation processes have to take the external resources into account.

Resource allocation processes that do not consider these external resources, and provide mechanisms to steer the resource allocation in these partner companies, and communities outside the boundaries of the focal firm, run the risk leading to misallocation of resources, and to the ultimate failure of the systemic innovation.

Chapter 4
ICT Tools and Systems Supporting Innovation in Product/Process Development

Abstract Information and communication technology (ICT) plays a key role in modern innovation and new product/process development. The chapter is dedicated to ICT tools supporting product/process innovation achieved throughout the development process. A general overview of such ICT tools is provided, specifically addressing Knowledge-based engineering systems and reasoning methods/tools, as well as tools to support innovation process in the Extended Enterprise context. Standardization aspects are of special relevance for application of ICT systems in industrial innovation processes. Due to their importance for the modern innovation processes, ICT to support collaborative product/process development and innovation and ontology management are addressed in more detail.

4.1 ICT Supporting Innovation in Product/Process Development

The new product/process development is tightly coupled with application of information and communication technologies (ICT) tools. It may be stated that ICT are a key enabler for innovation in modern product/process development in an Extended Enterprise (EE). ICT tools may facilitate innovation within product/process development in many different ways, *e.g.*, by supporting:

- Specific product/process design (design tools dedicated to specific topics)
- Knowledge management needed for product/process development
- Innovation in product/process development
- Innovation processes in EE
- Collaborative work within innovation processes
- *etc.*

Managing of knowledge needed for innovation within product/process development in general in an EE environment is a key issue, which in turn requires effective utilization of ICT. This section addresses primarily the application of ICT for *knowledge management* (KM) needed for innovation in product/process development in industry, but a brief overview of the tools for specific product/process design and tools to support innovation processes is provided. Architectural and standardization issues are briefly discussed as well.

4.1.1 ICT Tools Supporting Product/Process Design

There is a huge number of ICT tools supporting various specific product/process design processes. Two large groups will be briefly discussed here.

To the first group belong ICT supporting geometrical design of products – *computer aided design (CAD) systems.* It has been more than 35 years since the systematic approach to engineering design was firstly published. There is no doubt that since the late 1970s the way of carrying out engineering design has changed significantly from the technological and methodological point of view (see Chap. 1). From the technological point of view, the use of design support technologies such as CAD has become a day to day practice. However, systematic design was developed at the time when only big organizations, such as aerospace and automotive companies, had access to CAD technology. Even now some researchers recognize that, although CAD systems have had major impact on how design is accomplished in industry, the effects on the designers and on the final products are not sufficiently studied (Ullman 2002). The advance of the ICT to support design processes has been brought about by the deployment of three generations of CAD systems from the early geometric modelers to the current feature-based design systems (Horváth and Van der Vegte 2003). It has also been criticized that these technologies rely on information-based schemas rather than knowledge-based ones and that they do not support "cooperativity." For instance, Weber and Deubel (2003) argue that today's Product Data Management (PDM), Product Life cycle Management (PLM) and computer aided design (CAD) systems provide infrastructure to manage data, but do not retain knowledge about the content and the interrelationships of the data they handle. The CAD tools will not be specifically analyzed here, but it may be stated that current CAD tools are still missing a full provision of knowledge support.

The second ICT group to support product/process design represents *virtual reality.* Virtual reality (VR) is a technology with great potential in product/process design. The term "virtual reality" has been used by many researches and practitioners and may have different meanings. One of the best meanings is a method that allows people to visualize, handle, and interact with computers and extremely complex data. Virtual reality is an innovative interface. The main newness which is offered by this technology is its great potential regarding user interaction with computers, in comparison to classical tools. With these classical tools, the user could only act with mouse and keyboard to obtain 2D and 3D information in a traditional monitor, while virtual reality leads to the integration of the user as an element of the virtual environment.

In particular, industries from aerospace and automotive sectors are using virtual reality techniques. Nowadays, the use of virtual reality systems for evaluating design characteristics is fairly common and training simulators are being developed in many cases.[1]

Virtual prototyping allows reduction of time and costs of product/process development. New methods of virtual prototyping using techniques with high quality visualization and interaction can be applied in different areas such as automation, naval industry, or the aeronautical sector. Thus, virtual reality is a system that supports various industrial processes such as design, assembly, manufacturing, and human integrated design, while being compatible with current parametric CAD/CAM systems. The architecture behind such a system should allow for the expansion and customization of the virtual environments to suit the engineer's needs.

4.1.2 ICT Supporting Knowledge Management for Product/Process Innovation

Exploiting the capital associated with design knowledge has been shown to release considerable savings in the cost and lead times for design of new products.

The reason to seek knowledge support to engineering design may be found in a change in the engineering design and product development paradigm which is being introduced by the knowledge age, as explained in Chap. 3.

The step towards knowledge-based design technologies has not only been driven by the need of being in the "wave" of the so-called "knowledge age." This need has also been added by the request for considering additional design issues in *design support systems* (DSS), apart from only product geometry and information modeling. Knowledge-based infrastructure starts to be compulsory in DSS when one needs to extend the scope of computer aided design by adding the following elements:

- The need for support in the early stages of product design; this has been recognized as a critical issue concerning the development of DSS (see for instance Li *et al.* 2001; Takeda *et al.* 2003).

---

[1] For example, modern aircraft are so sophisticated and so expensive for real life use that simulators play an important part in design but also in pilots' coaching. In this way, simulators have extended their fields. They are used, for example, in astronaut coaching, in car systems validation, in testing the technical characteristics of new machines, *etc.*

- The need for handling "soft design requirements" such as those based on the environment; those based on the usability of products; or those based on the user intangible perceptions that boost the competitive advantage of products.
- The need for supporting knowledge management (KM) as a key concept to gain competitive advantage of enterprises in the long term. Until recently, engineering designers have been focusing their work just on the performance of their resulting artifacts. Nowadays the concerns are not only on how effective is the current product but also on how we can learn from the current product to provide better products in the future.

Though KM is one of the major key issues in a lot of business areas, it is still quite difficult to find many business implementations of KM for formalizing, structuring, and sharing product data/information. The main problems with the KM methods/tools are the re-use and sharing of knowledge among different partners within an EE, since most of the existing tools are missing capabilities to provide presentation of the captured knowledge in appropriate forms to different collaborating actors by provision of a personalized, context-sensitive, task-sensitive and role-sensitive functionality, and maintenance of such knowledge systems requires knowledge system specialists.

One of the key problems in application of tools to support KM in product/process development has to do with current engineering design practice. Despite the efforts that research and industrial community is putting on the development of wide-scope and generic design theories and design methodologies, as explained in Chaps. 1 and 2, the common perception is that engineering design industrial practice relies mostly on the traditional design frameworks, mostly represented by Pahl and Beitz (1996), and Ulrich and Eppinger (1995).

These design frameworks, while they have provided foundations and common understanding for engineering design practice, have also become "out of date" as a reference design modeling framework, leading to many obstacles in application of the ICT tools to support KM for product/process development. Some of the limitations of these approaches which are relevant here are:

- From the knowledge-based design perspective, if one takes a look at the design practice promoted by these traditional frameworks, guidelines to represent product knowledge acquired in the design process cannot be found. In this way, information is the only output expected. However, there is an enormous amount of knowledge that is generated by doing systematic design as it is stated by Pahl and Beitz (1996). This includes not only knowledge about physical artifacts, but also about processes required to produce them, knowledge about the organization, regulations, *etc.*
- On the other hand, most of the approaches to model the mentioned "soft design requirements" has come in the form of design methodologies that help to consider these requirements in the traditional design process. Design methods are usually domain-specific frameworks containing a "strict" description of

how to represent design problems and their solutions, which often leads to many problems in application of generic ICT tools.[2]

- The systematic approach also has a weakness in its strict approach to model design problems in its attempt to produce a deterministic view on how the design solution has to be reached. Such a lack of flexibility usually ends up in the need of many feedback loops that become especially time consuming as the design process reaches embodiment and detail stages.

In the search for competitive advantage drivers, the multidisciplinary research carried out in industry and academia has brought a number of design techniques and methodologies, as explained in Chap. 1. The mentioned QFD, FMEA, and VE are good examples of these methodologies, while feature-based design[3] can be identified as end-user technique. Furthermore, there have also been significant contributions to the way of thinking in respect of engineering design. An example is the introduction of the so-called "concurrent engineering" concept (see Chap. 3) which is one of the most significant advances regarding integration, which also considers the use of CAD tools (this will be analyzed in more detail in Sect. 4.2 and Chap. 5).

Apart from the problems related to these techniques, methodologies or wide scope frameworks, the integration of them in a consistent manner still remains an open issue. Engineering companies realize that the use of these techniques, methodologies or frameworks has direct implications in competitive advantage. What it is not completely realized is the "smooth" interaction between these techniques and their relationship with the design process itself.

The issue is especially critical when we consider ICT support to engineering design as a knowledge-centered activity rather than data or information-based task. Effective models of product knowledge should contribute to support decision-making activities in two directions:

1. Across different views of the design problem required to implement design methodologies
2. Across different stages of the product/process life cycle; from early definition of the problem; throughout the data generation that takes in the detail design; to the disposal and recycling stages

The development of ontologies to unify and to put into context the different concepts and terms of the industrial domain can be very helpful to avoid misinter-

---

2 These design methods tend to be "*ad hoc*" recommended practices to address problems that are relevant for different points of view of the engineering design activity. Examples of these methods are QFD (Quality Function Deployment), FMEA (Failure Mode and Effect Analysis), VE (Value Engineering), *etc.* This is a collection of methods that engineering designers adapt to their needs and they are usually built on the top of the theory-based frameworks.

3 Feature-based design is an approach that promises substantial benefits in capturing and transferring a designer's intent to other stages of design and manufacturing, thereby helping to integrate the diverse functions associated with design and manufacturing (O'Grady and Kim 1996).

pretations. Therefore, application of ontologies is promising to solve basic problems of sharing of knowledge and its importance is being recognized in many research fields and application areas, including knowledge engineering, database design and integration, information retrieval and integration. Ontologies play an important role in KM in modern product/process development and innovation. Therefore, a separate Sect. 4.3 is dedicated to ontology aspects in innovation.

Some practical examples of ICT tools to support KM in collaborative innovation process are described in Chap. 5.

4.1.2.1 Knowledge-based Engineering Systems

Knowledge-based engineering (KBE) concerns the computerization of processes associated with industrial products design, usually routine design. KBE applications are intended to capture and codify the domain expertise of contributors throughout an organization. The design problems and knowledge are translated into sets of rules.[4] Until now there has been no common way of collecting, structuring, and formalizing the engineering knowledge associated with designs. This not only makes it hard to plan and organize the process of building KBE applications but also means that updating them and re-using modules implies a number of problems.

KBE systems are effective in automated detailed design using engineering rules. However, they are generally limited to very routine and small subassemblies of products or traditional products. For such small subassemblies both product and design process are often constrained and KBE systems can be effectively applied. Moreover, KBE systems are dedicated to the use of one specific expert engineer and require a KBE expert to be configured. Their use within collaborative work specifically implies many difficulties.

4.1.2.2 Reasoning Methods

Reasoning methods and tools provide ways to use existing knowledge for reasoning purposes. They are used for KM in industry related to different topics and applications (*e.g.*, diagnostics, maintenance, *etc.*). Generally, the complete life cycle of KM is supported. These tools provide a means to:

- Capture knowledge in different forms
- Store the knowledge

---

[4] There are several software packages for the development of specific KBE applications. Some interesting examples are: Adaptive Modeling Language (AML) from TechnoSoft, (http://www.technosoft.com/), Design ++ from Design Power (http://www.dp.com/), *etc.* Note: Adaptive Modeling Language and AML are copyright of TechnoSoft, Inc., 11180 Reed Hartman Highway, Cincinnati, OH 45242, U.S.A.

- Provide mechanisms to re-use this knowledge for reasoning on "similar" problems including:
  - Retrieving
  - Identification of similarity
  - Analyzing
  - Creating "new" solution
- Maintenance of knowledge

Therefore, under reasoning methods and tools it is possible to group the set of algorithms/procedures/tools for KM using different approaches to support reasoning on different topics/problems. Consequently, they are also applied to support product/process development. For example, they may be applied to search for "similar" previous design which may be re-used, *etc.* (see Sect. 5.3.2.4).

There are many commercially available products providing means to capture knowledge on problems and product/processes. Such tools primarily address heuristic reasoning, *e.g.,* case-based reasoning (CBR), rule-based reasoning (RBR), and model-based reasoning approaches.[5]

These generic tools normally need "heavy" customization and adaptation to be applied to product/process development and innovation. One of the main problems with these reasoning methods/tools is a re-use and sharing of knowledge among different experts and partners within distributed and extended industrial companies, since most of the existing tools are missing capabilities to provide presentation of the captured knowledge in appropriate form to different actors.

4.1.3 ICT Tools Supporting Innovation Process

Several methods to support innovation process and generation of ideas are analyzed in Chap. 1. Most of them are supported by certain ICT tools. In this section several examples of such tools are presented. An example of powerful system to support collaborative innovation process is presented in more detail in Sect. 5.3.

---

[5] Some examples are listed here: good examples of CBR tools that were used in the past for many applications are eGain KnowledgeAgent™, Easy Reasoner, Know How. eGain KnowledgeAgent™ is a trademark of eGain Communications, 345 E. Middlefield Road, Mountain View, CA 94043, U.S.A. http://www.egain.com/.
They all are tool kits to develop case-based reasoning applications. For example, eGain KnowledgeAgent is a KM toolset, comprising: eGain Knowledge Gateway, eGain KnowledgeAgent, and eGain Knowledge Self-Service (http://www.egain.com/solutions/help_desks.asp).
HELPDESK-3 is a CBR tool designed to automated help desk, etc.
Several free software tools for CBR and RBR are available, *e.g.,*
- Jcolibri (http://www.jcolibri.com),
- Jess (http://www.jessrules.com),
- FuzzyJ Toolkit (http://www.iit.nrc.ca/IR_public/fuzzy/fuzzyJToolkit2.html),
- JBossRules (http://www.jboss.com/products/rules), *etc.*

Tools supporting creativity. There are many software tools based in the TRIZ methodology (see Sect. 1.4.) for inventive problem solving. Some typical examples are:

- *IWB (Innovation WorkBench®)*[6]
- *TECH OPTIMIZER*

Both software packages (both based in USA) use schematic representation of problems and automated analysis of generated diagrams that guides the user to the abstract solution. Technical information and examples are included for helping the user in the particularization of the solution. However, these are both primarily aimed at the scientist level of user, and not at the industrial manufacturing level.

Another example of TRIZ-based tool is *Southbeach*[7] which uses typical TRIZ notation as a new visual modeling and diagram style.

Another tool for helping to generate ideas *Ideafisher,*[8] Inspiration professional edition. However, again this is aimed at the specialist level, and may not be appropriate for industrial companies.[9]

Methodology and tool known as "Logo Visual Technology" (LVT)[10] support the organization and delivery of knowledge for innovation. The tool has several important features. The first one is a dimension of *tangibility* and the second one a way of handling knowledge in terms of units that belong to a higher level than words, and constitute whole "molecules" of meaning (MM) in their own right. This theory is strongly based on the humanization of systems of thought, which must allow for creativity and change.

Another theory, called "MindMaps", has been formulated for solving the same KM and transference problem. MindMaps®[11] are a concise way of displaying notes and information and their associations. Tony Buzan developed Mind Maps as an efficient way of using the brain's ability for association (Buzan 2003). Association plays a dominant role in nearly every mental function, and words themselves are no exception. Every single word and idea has numerous links attaching it to other ideas and concepts.

---

[6] Innovation WorkBench® is a registered trademark of Ideation International Inc., 32000 Northwestern Hwy, Ste 145, Farmington Hills, MI 48334, U.S.A. http://www.ideationtriz.com

[7] http://www.southbeachinc.com/

[8] http://www.thoughtoffice.com/?page_id=148

[9] Some tools exist to support application of other methodologies: QFDCapture™ from ITT for QFD (http://www.qfdcapture.com), GAMDEC/GAMTREE for FMEA (http://www.gamtech.fr/Fiche%20FIABILITY%20GB.pdf), EXPERT CHOICE supporting decision making (http://www.expertchoice.com), *etc.*

[10] http://www.logovisual.com

[11] Mind Map® is a registered trademark of The Buzan Organisation, Harleyford Manor Estate, Henley Road, Marlow, Buckinghamshire SL7 2DX, United Kingdom http://www.buzanworld.com

Tools supporting gathering of ideas. Many companies are using their own tools to collect and manage ideas. Here, two typical examples will be briefly presented.

Siemens' 3i. Siemens has an idea management program, called 3i (Ideas, Impulses and Initiatives) which the company encourages its employees with to take the initiative by suggesting and implementing improvements (Siemens 2002).

Employees are invited, as part of the 3i Program, to act on their own initiative to develop and implement suggestions for improvements within the company. Suggestions relating to work safety and operational environment protection are of particular importance. Managers are invited to encourage and support employees. They should promote work rising from group initiative and provide the employees involved in such work with the necessary means.

Chrysler's EBOK. The problem at Chrysler that resulted in the development of the Engineering Book of Knowledge (EBOK) is the common problem of many enterprises: the inability to transfer knowledge within the enterprise. The objective of the Engineering Book of Knowledge (EBOK) was to share engineering knowledge across divisions, avoid duplicate work, and ensure reuse. EBOK is not an ordinary database; it is not only accessible, useful, and comprehensive, it also consists of people's opinions, best practices, and similar things learned simply through experience (Turban *et al.* 2006)

The EBOK was built on top of Lotus Notes®[12] technology, with the book metaphor used to organize the knowledge and provide user-friendly navigation. GrapeVINE, a Notes add-on application from grapeVINE Technologies, is used for monitoring and classifying information from the Lotus Notes® databases. It also alerts users when relevant documents, based on the user's interest profile, are added or changed.

Chrysler has improved engineering efficiency by reusing knowledge and avoiding duplication of effort, thus shortening time to market. Improvements in quality have also been realized because of the capability to share judgments and engineering results. There are no objective "measures" of results to determine success or failure of the EBOK, but qualitative reports are providing adequate justification for the knowledge representation project. Additionally, the use and re-use of best practice engineering projects or developments have much more visibility and provide a sense of community for those who apply it (Bair 1997).

Tools for innovation assessment. Several tools exist to help assessing innovation capacity. Some typical examples are:

---

[12] Lotus Notes® is a registered trademark of International Business Machines Corporation in the United States, other countries, or both.

- *Innovation Styles*®[13] focuses on evaluating *how are you innovative* rather than *how innovative are you.* It is based on the fact that all people are unique individuals and while everyone has the capacity to be creative and innovative, each of us expresses this potential differently
- *The Innovation Assessment Program* by United Inventors Association[14] is an inventor/innovator assistance service that provides inventors, entrepreneurs, and product marketing/manufacturing enterprises with an honest and objective third-party analysis of the risks and potential of their ideas and inventions. This is why it focuses on invention evaluation.

Several packages have modules that practice systematic assessments. The assessment of ideas and innovations must be carried out following certain criteria, and has to be specific for the business area and process/product of the user and development team. Therefore, standard tools are generally clustered, based on the working area, and organized to obtain maximum efficiency.

Decision trees are excellent tools for helping developers of innovative solutions to choose between several courses of action. They provide a highly effective structure within which the user can lay out options and investigate the possible outcomes of choosing those options. They also help the user to form a balanced picture of the risks and rewards associated with each possible course of action. Here they can be used to make a first assessment of the potential solution, checking its fitness with the firm policy and resources.

Decision trees provide an effective method of decision making because they:

- Clearly lay out the problem so that all options can be challenged
- Allow one to analyze fully the possible consequences of a decision
- Provide a framework to quantify the values of outcomes and the probabilities of achieving them
- Help to make the best decisions on the basis of existing information and best guesses

Reasoning tools. As will be explained in Chap. 5, reasoning tools, such as RBR or CBR, can also be used to assess and propose the level of priority of each idea as the rules can be established by gathering information about the business objectives and users' satisfaction.

[13] Innovation Styles® is a registered trademark of Innovation Styles Inc., 150 East 18th Street, Suite #2E, New York, NY 10003, U.S.A. http://innovationstyles.com.
[14] http://www.uiausa.org/Default.aspx?page=129

4.1.4 ICT Architectures to Support Product/Process Development and Standardization Aspects

As ICT tools to support innovation processes in industry often have to be combined and they have to facilitate collaborative work of many actors within an EE, standardization and ICT architectural aspects play a most important role in modern industry.

Different activities were launched to provide access to central consolidated data repositories describing a product/process. Within product/process design, the product information is stored in different systems like:

- *Product Data Management (PDM)*
- *Computer aided design (CAD)*
- *Computer aided manufacturing (CAM)*
- *Enterprise resource planning (ERP)*
- *Engineering data management (EDM)*
- *etc.*

These product/process descriptions and their corresponding models are mostly used as standalone versions.

The demand of *interoperability* between different systems used in different phases of the product/process life cycle and by different stakeholders leads to an exponentially increasing number of tool-to-tool interfaces. Large companies and increasing numbers of small and medium sized enterprises (SME) are now on the way to implement neutral interfaces for the exchange of their products information in order to:

- Reduce:
  - Implementation and life cycle costs
  - The product/process time-to-market
- Increase flexibility and agility

The creation of product or even system information models to formalize and structure product/process information is a very important task.[15] During the last few years several reference models have been developed within different standardization activities. Reference models on various levels of abstraction can be relevant to any industry, business area or company. Examples for reference models are, for instance, the IAA (IBM Insurance Application Architecture), the DZ-SIMPROLOG (Reference Models of Fraunhofer), the ODP (Open Distributed Processing), and OMA (Object Management Architecture).

[15] United Nations Standard Product and Services Classification (UNSPSC) Code organization. http://www.unspsc.org/home.htm

Many standards highly relevant for ICT supporting product/process development exist or are emerging. The most applied international standard for the representation and exchange of product data/information is STEP, which is an international standard (ISO 10303) since 1994. STEP represents a viable alternative to the current chaos of multiple and fragmented standards and data forms. One of the important Application Protocol developed within the ISO working group and the SEDRES-2 project is "AP233 Systems Engineering data representation." The fact that systems engineering incorporates a lot of different domains through the whole product life cycle makes it an excellent baseline for representing data of complex products in a complete and consistent way.

Several generic *ICT architectures* are developed and widely applied in industry and many of them are applied to support innovation processes as well. Here, the emerging architecture which is likely to be specifically appropriate for collaborative innovation: *Service Oriented Architecture (SOA)* will be briefly described (see Sect. 4.2 and Chap. 5) instead of analyzing different ICT architectures, as the latter is out of the scope of this book.

SOA[16] represents a logical way of structuring a software system into a set of loosely coupled components whose interfaces can be described, published, discovered, and invoked over a network. These components are deployed as services with standardized interfaces, independent of any specific platform or implementation technology (thus separating implementation aspects from interaction specifications) and that carry out together a high-level function or business process, *e.g.*, placing an order, making a credit approval on a purchase (Papazoglou and van den Heuvel 2007). The Organization for the Advancement of Structured Information Standards (OASIS) specifies a service in the scope of SOA as "a mechanism to enable access to one or more capabilities, where the access is provided using a prescribed interface and is exercised consistent with constraints and policies as specified by the service description."

[16] The SOA paradigm is based on the broker architectural pattern, where a central component of the architecture (the service broker) acts as an intermediary between the other interacting components (service provider and service consumer). In SOA the broker acts as a mere searchable repository of service descriptions (interfaces) where service providers publish their services and service consumers find services and obtain binding information for these services. Hence, the broker in the SOA model can be regarded as services yellow pages (analogous to the telephone yellow pages) having a passive role, *i.e.*, not interfering with actual interaction between the consumer and the provider. The service is explicitly declared and published by the component that offers it. The service consumer is responsible for starting the interaction with the service provider after discovering the desired service contract (interface) published in the broker and performing a binding process to the service provider. Once the interaction is started, the execution is autonomous and cannot be influenced by any external inputs, including the broker.

Web services are currently the most promising and most widely used technology that enables realizing the principles of SOA. They represent self-contained modular business applications that have open, Internet-oriented, standards-based interfaces. The *World Wide Web Consortium* (W3C) defines a Web service as "applications identified by a Uniform Resource Identifier (URI), whose interfaces and bindings are capable of being defined, described and discovered as *eXtensible Markup Language (XML)* artifacts. A Web service supports direct interactions with other software agents using XML-based messages exchanged *via* Internet-based protocols."[17]

In this sense, they provide the basis for the development and execution of business processes that are distributed over the network and available *via* standard interfaces and protocols. Thus, Web services enable the integration and interoperability of heterogeneous systems and components which may be geographically dispersed. The vision behind the technology is to transform the Internet into an environment where businesses can expose their current and future business applications as Web services that can be easily discovered and used by interested parties.

To realize this vision, the Web service concept relies on a widely available and strongly supported set of standards and protocols. It can be said that the widespread adoption of Web service standards has reinvigorated this approach by providing a universally accepted set of interoperability standards for building, describing, cataloguing, and managing reusable services, *e.g.*, XML, *Simple Object Access Protocol (SOAP), Universal Description Discovery and Integration (UDDI), Web Service Description Language (WSDL), Web Services Interoperability Organization (WS-I)* standards, *etc.* This is a distinct advantage over other architectures, since it makes interoperability one of its intrinsic characteristics, which eases the integrations of heterogeneous systems and provides a major enhancement in business agility (van der Meer *et al.* 2006). One of the main challenges for the systems is effectively to organize and discover their services in a uniform, interoperable manner, so that the customization, reutilization, and integration of existing services are effective.

4.2 Collaborative Working Environments for Innovation in Product/Process Development

The product/process development and innovation processes in general in modern industry are increasingly collaborative processes: they require collaboration of many teams within an Extended Enterprise (EE) context, where teams are often geographically distributed. Such collaborative work requires support by ICT solu-

---

[17] W3C Homepage: http://www.w3.org.

tions. As explained in Sect. 3.3, collaborative work is a dominant approach for innovation in product/process development in the twenty-first century.

Collaboration technologies are the key to collaborative design industrial environments that enable people (including designers, engineers, managers, and customers) to collaborate and interact on the development of a new product/process regardless of their geographic locations and interaction means. *Collaborative Working Environments (CWE)* are intended to assist users to participate in different workspaces, discover collaboration opportunities, provide knowledge or services, seek information or assistance, and perform and coordinate their product and process development activities.

Collaboration@Work[18] addresses different areas of collaboration: industry, science/research, medicine, government, *etc.* Collaboration technologies are and will be the key technologies for product and process design as well as enterprise integration and collaboration. As jobs and factories are distributed around the globe, real-time information technology will be the most effective means of collaboration.

Commercially, the products that support collaborative engineering typically are grouped into three categories: groupware, teamware, and taskware.

- *Groupware.* It includes such things as e-mail and communication, negotiation, and meeting support software
- *Teamware.* It supports teams interacting in product development. Typically, it is embedded within a process model and provides a capacity to share work products. Formal design methods and techniques (see Sect. 1.4) such as Quality Function Deployment (QFD), Value Analysis (VA), and Failure Mode and Effect Analysis (FMEA) are often used as team tools. Team members can use standard browsers to access these services at different locations. In comparison with standalone computerized design tools, Web - based systems are effective in facilitating teamwork in product/process development
- *Taskware.* It focuses on particular tasks, and typically cannot be shared across tasks

During the past few years, the Web-based infrastructure has also been used in developing collaboration systems for product/process data sharing and exchange, manufacturing process monitoring and control and particularly for enterprise (business) integration, enterprise collaboration and supply chain management (see Chap. 5).

This section provides a definition of CWE and an overview of the current solutions for CWE. The special emphasis is put upon CWE for innovation in industry, as well as upon standardization aspects and security aspects. The practical applications of CWE for product/process development and innovation processes are described in Chap. 5.

---

[18] The expression Collaboration@Work is extensively used to name ICT tools to support collaborative work (Laso-Ballesteros and Salmelin 2005).

4.2.1 Definition

Generally speaking, CWE is a set of tools to support collaborative work. It is usually assumed that these tools are ICT-based, but they may include different means as discussed in Chap. 2: architectural and spatial facilities, furniture, *etc.*, to support collaborative work. In the text to follow the focus will be upon ICT tools and under CWE will be understood ICT environment to support collaborative work.

eCollaboration or Collaboration@Work means collaboration among individuals engaged in a common task to achieve a shared objective using CWE technologies (UNIT F4 2005).

The current CWE only partly fulfils the industrial needs. The vision is (Expert Group 2004): "*Next Generation Collaborative Working Environments will deliver quality of experience to co-workers, will be based on flexible services components and customized to different communities.*"

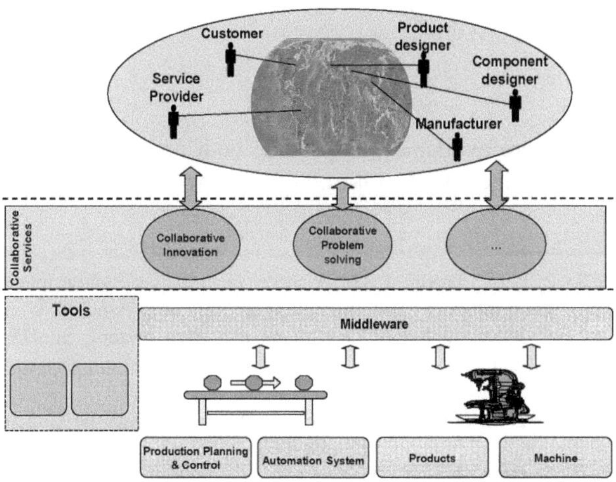

Fig. 4.1 Collaborative Working Environment supporting geographically distributed teams

Figure 4.1 represents a very generic architecture for CWE in (manufacturing) industry which can be instantiated for different Extended Enterprise environments, applications, *etc.* Practically all CWE discussed in Chap. 5 follow this generic architecture.

CWE intend to support collaborative work in both physical and virtual space. Therefore, the paradigm of "hybrid" virtual-real working environment is emerging. A "hybrid" virtual-real environment is an optimal infrastructure for creative collaborative work. Virtualization of products/processes allows for effective collaborative analysis of new ideas, experimenting to test different ideas, collaborative problem solving, optimal distribution of work on innovation, *etc.* A hybrid virtual industrial environment has to enable effective collaborative work among geographically distributed teams, as well as teams involved in different life cycle phases of products and processes (teams distributed in time), and also among industrial teams and wider communities, as it may provide deeper and more effective assess to products/processes.

4.2.2 Overview of Needs and Approaches/Tools

CWE aim at improving human abilities to work collaboratively, thereby increasing creativity, which in turn, boosts innovation and productivity as well as support new value creation forms. Generic CWE solutions are already applied in many domains and applications such as science, education, public domain, *etc.,* but many research issues are still open. This section provides a brief overview of key issues relevant for CWE in general, the needs and current approaches/tools.

There are many tools to support collaborative work and some of the relevant systems for eCollaboration include different collaboration platforms such as:

- Basic Support for Collaborative Work[19]
- Isabel Application[20]
- Marratech[21]

A number of commercial tools are available that provide both product life-time management and collaborative engineering facilities. Some examples are:

---

[19] The BSCW server is written in Python®. To run a BSCW server one needs a Python interpreter for his machine (including the standard Python® libraries) as well as a standard Web server. The BSCW server runs on UNIX (including the Solaris operating system, SunOS, Linux®, DEC OSF, HP-UX, Irix®, BSD/OS and AIX®) and Windows NT® operating system (version 4.0 or above).
Windows NT and Windows are registered trademarks of Microsoft Corporation in the United States and other countries.
UNIX is a registered trademark of The Open Group.
Sun, Sun Microsystems, the Sun Logo, Java, SunOS, Solaris and Enterprise JavaBeans are trademarks or registered trademarks of Sun Microsystems, Inc. in the United States and other countries. Worldwide Headquarters, Sun Microsystems, Inc., 4150 Network Circle, Santa Clara, CA 95054, U.S.A. http://www.sun.com/suntrademarks.
"Python" and the Python logos are trademarks or registered trademarks of the Python Software Foundation, used with permission.
IRIX® is a registered trademark of Silicon Graphics, Inc., in the United States and/or other countries worldwide.
AIX is a registered trademark of International Business Machines Corporation in the United States, other countries, or both.

[20] The Isabel CSCW application (http://www.agora-2000.com/products/isabel/) is a group collaboration tool for the Internet. Isabel supports the realisation of distributed meetings, classrooms, congresses, *etc.* (Quemada *et al.* 2005).

[21] The Marratech client is freely available software that is easily installed. Marratech gives an access to a secure group work environment with voice over IP, an interactive whiteboard, the ability to share information and documents, talk and chat in groups or in private and, if desired, the opportunity to see each other by using Web cameras.

- *SMARTEAM V5* from IBM and SMARTEAM, supports product life cycle management and collaboration across the EE[22]
- *ENOVIA* supports product knowledge management for the entire product life cycle from initial concept to product in service, as well as collaboration and teamwork[23]
- *MatrixOne*: Web-based design collaboration and product-development software tools[24]
- Windchill®[25]
- NX[26] support product design in the EE[27].
- SAP®[28] provides an integrated product life cycle management solution for managing the complete product life cycle of the extended supply chain, from design and production through sales and maintenance[29]

A number of big organizations have also developed their own collaborative design systems, either as pilots scale or for full-scale application.[30]

However, many of the available systems for collaborative work are still used within the academic community. Those available on the market are oriented towards very big organizations – they are often very expensive and cannot be used by SME. Even large companies may encounter various problems when developing and introducing CWE for their specific needs. For example, recently, the Automotive Industry Action Group's Collaborative Conferencing Work Group has established data collaboration requirements, reviewed various solutions from about 40 vendors, and concluded that "while all vendors were able to address some of the requirements, no single vendor was able to meet all the requirements".

In fact, most current collaboration tools and environments provide a set of persistent services to users. However, they often rely on a centralized infrastructure. This makes the tools impossible to use when a specific resource or server is unavailable. Ideally, the collaboration environment should not depend on any specific resource or server; instead, the resources and servers should add value to the system when they are present. In addition, this infrastructure-centric approach

---

[22] http://www-01.ibm.com/software/plm/de/products/smarteam_latest.html

[23] http://www.3ds.com/products/enovia

[24] http://www.matrixone.com

[25] http://www.ptc.com/products/windchill. Windchill® is a registered trademark of Parametric Technology Corporation, 140 Kendrick Street Needham, MA 02494, USA http://www.ptc.com/"

[26] NX is a trademark of Siemens AG, Wittelsbacherplatz 2, D-80333 Munich, Germany http://www.siemens.com

[27] *e.g.,* http://www.plm.automation.siemens.com/en_us/products/nx/index.shtml

[28] SAP is the trademark or registered trademark of SAP AG in Germany and in several other countries.

[29] http://www.sap.com/index.epx

[30] An example is the DESIGN CONSULTANT system developed by the Knowledge – based Engineering Department of Ford Motor Company (Visteon).

makes these tools difficult to set up and scale, particularly when security is involved.

A collaboration environment should be structured to support informal, spontaneous collaborations as well as highly structured environments and it should allow for "the secure real-time data collaboration from any software program, used on any industry standard hardware platform, to virtually anywhere in the world."

Organizations constantly search for innovative applications and services to improve their business processes and to enrich CWE of their distributed and mobile knowledge workers. It is increasingly becoming apparent that a limiting factor in the support of more flexible work practices offered by systems today lies in their inherent assumptions about (Expert Group 2006):

1. Technical infrastructures in place (hardware, software, and communication networks)
2. Interaction patterns of the users involved in the processes

Emerging new ways of flexible and mobile teamwork on one hand and dynamic and highly agile (virtual business) communities on the other require new technical as well as organizational support, which current technologies and infrastructures do not cater for sufficiently.

Pervasiveness of collaboration services is an important means in such a context to support new business models and encourage new ways of working. Recent developments show a strong move towards increasingly mobile and virtual project teams. Whereas traditional organizational structures in business relied on teams of collaborators dedicated to a specific project for a long period, many organizations increasingly rely on nimble teams, formed from members of possibly different branches or companies, assigned to perform short-lived tasks in an *ad hoc* manner (sometimes called *ad hoc* teams). Consequently, the emerging new styles of distributed and mobile collaboration are fostering new interaction patterns of working. Interaction patterns consist of information related to synchronous and asynchronous communication on one hand and the coordination aspects on the other hand (for more details on collaboration patterns see Chap. 5).

Collaboration@Work has at least four dimensions (Expert Group 2004):

- Users and group members (co-workers)
- Working processes
- Technologies
- Application areas

CWE technologies have to integrate all these dimensions into a suitable collaboration at work platform and will have to provide support and openness for all mentioned aspects as explained below. From a user-centric point of view, CWE have to deliver "Quality of Experiences." The focus of a co-worker lies on her/his primary tasks. The co-worker needs a seamless integration of ICT technologies into his/her work and a natural, unobtrusive and "causal" interaction with the used devices and technologies. Collaborative technologies have to be able to define the available services that will be required for fulfilling the co-worker's tasks. CWE technologies have to provide seamless integration of synchronous and asynchronous communications and maintain the user experience in both connected and disconnected modes. They have to support the bundling of different (mobile) devices, and they have to be able to orchestrate the services into the application that will be delivered to the workers. They have to solve the needs of:

- Accountability
- Understandability
- Tangibility
- Awareness of an integrity level

The so-called *WEB 2.0* principles are of key importance for CWE. Web 2.0 principles promote service building and composition on individual level. It is expected that future developments would continue the move from the service composition at individual level of Web 2.0 to company and group level – thus moving from "mySpace" to "mySME" – (see Expert Group 2006). There is a clear trend to seek applications outside of individuals and small groups to organizations and enterprises. It is also likely that machines will be included as a part of own extended information and collaboration network and that collaboration will take place on a massive scale and global geographical mode.

The emphasis is on services and tools for collaboration design, *i.e.,* tools which either off-line or on-line support design of collaboration, development of core functions exposed as services and their:

- Composition
- Customization
- Update
- Maintenance, *etc.*

The objective is "to contribute to blurring the boundary between designing and using, addressing implicitly or explicitly Web 2.0 aspects" (Expert Group 2006). By this the ultimate goal of building collaboration services quicker, cheaper, more flexible, with higher quality, will be achieved. The objective is also to build reference models for CWE that support "automatic" composition and integration of basic functions exposed as services, taking into account Web 2.0 paradigm. Therefore, modern CWE are aiming to provide means for industry, including SME, to effectively build/compose services and trying to solve critical problems related to moving from "mySpace" to "mySME," such as composition of services taking into account IPR and privacy issues, sharing of knowledge objects, *etc.*

From a technique-centric point of view, CWE have to be based on flexible service components allowing adaptability and scalability (Prinz 2005). They have to be interoperable both at a syntactic and semantic level, and also at the protocol level. Therefore, standardization at a semantic level is necessary (see Sect. 4.2.4).

4.2.3 eCollaboration for Innovation in Industry

This section is focused upon specific requirements on CWE in industry, and specifically upon requirements on CWE for collaborative innovation processes, *i.e.*, for *e-Innovation.*

As explained before (see Chap. 3), modern innovation processes in industry require new forms of collaboration between the various teams involved in a product/process innovation (*e.g.*, design, planning, production scheduling, manufacturing, after sales services, *etc.*). Especially collaboration amongst different actors in innovation processes in industry is currently burdened with a number of problems concerning distribution of work, synchronization and persistence of workspaces, *etc.* The problem is the teams in modern and highly flexible industry require often different collaboration patterns, *e.g.*, a combination of synchronous and asynchronous collaboration during innovation processes (Stokic 2006). Collaboration for decision making purposes within innovation and concurrent engineering processes has to integrate effective information sharing and activity synchronization (Miao and Haake 1998).

Work and resources sharing within an enterprise (*e.g.*, amongst different areas, departments, plants, different players in a virtual company, *etc.*) and application of ICT tools to support collaborative work on innovation are related to a number of problems such as how to optimally share and activate knowledge and distribute work among teams from human perspective. Existing, often powerful methods and ICT tools nevertheless do not satisfy many of these requirements. These problems are often consequence of inadequate information on actual circumstances/context within which different actors operate (environment and contextual awareness) and on their actual capabilities, as well as an inappropriate interaction amongst different actors when making decisions on distribution and synchronization of work (*e.g.*, when activating knowledge). This is the reason why collaboration amongst teams is still frequently managed using insufficiently systematic methodologies or following non-human centered concepts.[31] This specifically may constrain collective creativity and collaborative work on innovation in industry.

Innovation teams and communities impose requirements on collaborative services, such as:

- Accessibility to collaborative services at any time, from anywhere and from any device
- Pervasiveness of collaborative services in the sense that they support new business models and encourage new ways of working and novel interaction patterns (UNIT-F4 2004, UNIT-F4 2005)

[31] The problem often is how efficiently and promptly (on-line) to acquire and provide information/knowledge needed for such collaboration (*e.g.*, difference in work synchronization in shopfloor and design office, *etc.*) and how to use effectively such knowledge to support decisions regarding collaborative work (to distribute and synchronise the work).

Collaborative services shall be

- Dynamic
- Flexible
- Adaptable to fit any form for collaboration

Industrial companies also require special considerations regarding the way the actors of the EE work, *i.e.,* the collaboration patterns. Manufacturing environments provide a wide set of possible patterns, where the characteristics of the involved actors vary immensely. The main collaboration patterns in industry are identified in "SAMS: Synchronous, Asynchronous, Multi-synchronous Environments" (Molli *et al.* 2002) and discussed in Chap. 5.

The application of CWE in industry, specifically for collaborative innovation and engineering, has been subject of many research activities. Different algorithms for *e.g.,* collaborative computer-aided design (Li *et al.* 2006) and various ICT tools to support collaborative work (Drira *et al.* 2001; Churchill *et al.* 2001) were proposed. Besides generic tools for collaboration (e-mails, "blogs and wikis") widely applied in different communities, specific solutions for industry have also been developed. However, as indicated above, collaboration among teams with different technical backgrounds and with different collaboration patterns (*e.g.,* shop-floor teams and design teams, organized teams, and *ad hoc* communities, *etc.*) has not been sufficiently explored, and ICT solutions supporting such collaboration are generally missing. Although the prime objective of CWE in industry is to focus on organized teams, the manufacturing industry has to be open (on longer-term) for collaboration with a wider community. Subsequently, pro-active, culturally-aware services allowing access to virtualized resources and knowledge to support creativity by involving teams within wider communities in generation of new ideas on products and processes and solving problems are needed.

Many of the addressed problems are common for collaboration work in many different domains (as explained in previous section). However, there are several specific issues related to CWE in industry which impose the needs for RTD activities specifically focused upon (manufacturing) industry. From what is said above, such issues can be summarized as:

- High difference in working environments of the teams that have to collaborate, *e.g.,* shop-floor, logistics area, office area for design teams, *etc.*
- Different technical backgrounds of teams collaborating on a common problem, *e.g.,* process improvement requires collaboration of shop-floor workers with high practical experience but (often) low technical backgrounds, designers with high technical expertise, *etc.*
- (On-line) information/knowledge provision is usually more complex and critical than in other domains
- Combination of physical/virtual collaboration (hybrid collaboration) is more dynamic than in other domains
- Specific security and IPR requirements

- Strong focusing on organized groups but there are evident needs to include more *ad hoc* groups in collaborative work (specifically regarding innovation)

Table 4.1 Problems and possible solution approaches for CWE in industry (Stokic 2006a)

Topic	Problem	Possible solution approaches
Tools to support specific collaborative work in industry	Definition of collaborative patterns in (manufacturing) industry	Analysis of collaborative patterns, definition of collaboration models for (manufacturing) industry
	Generic services satisfying different collaborative patterns	Core collaborative services supporting different collaborative patterns
	Tools for design of collaboration services by non-IT experts	Tools for generation/update of collaborative services
Information infrastructure for collaboration	Information infrastructure to meet needs for collaboration within EE (including virtual product/processes)	Information middleware for collaboration, virtualized resources
Collaborative multi-criteria decision making	Work on traceability of multi-criteria decision making in industry	Multi-criteria decision making in collaborative environments, with main emphasis on responsibility of the decision, traceability, selection and weight of the decision criteria, representation and handling of uncertainty.
Security and IPR issues	How to share knowledge effectively among industrial teams within EE and with the wider communities?	Services/tools for CWE in industry, taking into account access capabilities in industrial companies
Semantic-based knowledge management (KM) for collaborative work	How collaborative work may support KM in industry? How to adapt knowledge management to different technical backgrounds?	Monitoring of work of individuals and groups to predict and adapt KM tools to people. Tools to adapt the "same knowledge" to people with different backgrounds
Collaboration among organized and *ad hoc* teams	Tools to support collaboration among teams within EE and wider communities	Pro-active, cultural aware services allowing access to virtualized resources and knowledge to support creativity by involving teams within wider communities in generation of new ideas on products and processes and solving problems
CWE architecture	Reference architecture for collaborative work in (manufacturing) industry (see Sect. 4.2.4)	The emerging reference architecture for application in industry – addressing smooth "upper layer" middleware interaction with the underlying layers

Table 4.1 provides an overview of the identified gaps between the available solutions and industrial needs and indicates possible approaches to solve these problems. These solutions are specifically elaborated in Chap. 5.

4.2.4 Standardization Aspects for Collaborative Working Environments

Standardization aspects are especially important for ICT solutions to support collaborative work. A number of standards related to different collaborative work aspects are emerging. The standards relevant for CWE are mostly related to

- Semantic-based KM
- Reference architectures

Standards related to semantic-based KM. As already mentioned in Sect. 4.1.4, of particular interest are grounding standards such as *XML*, *Resource Description Framework (RDF)* and *Web ontology language (OWL)*, set by W3C. XML and RDF are already widely adopted by industry, so their usage will allow CWE to work seamlessly with many systems. OWL industrial adopting is currently in the process of finalization.

As data exchange within EE plays a key role, the standards regarding interoperability are of high importance. Among these, special emphasis should be put on:

- *ebXML* (e-business XML), an open XML-based infrastructure enabling the global use of e-business information in an interoperable, secure and consistent way[32]
- *UN/CEFACT*[33] (United Nations Centre for Trade Facilitation and Electronic Business) has developed and promoted many tools for the facilitation of global business processes including *UN/EDIFACT* (United Nations Electronic Data Interchange For Administration, Commercc and Transport), the international EDI standard. Its current work program includes Simpl-edi and Object Oriented edi
- *OASIS*, as mentioned before, is the international, non-profit consortium that advances e-business by promoting open, collaborative development of interoperability specifications (OASIS 2006). Of major interest for CWE solutions in industry is the OASIS Standard Web Services Business Process Execution Language Version 2.0 (WS-BPEL 2.0)

Another relevant standard is the *Open System Architecture for Enterprise Application Integration* (*OSA-EAI*) specification, developed by the *Machinery In-*

---

[32] http://www.ebxml.org

[33] http://www.unece.org/cefact

formation Management Open Systems Alliance (MIMOSA). The MIMOSA-OPC Group has published its initial Open O&M Information Standards which combine the OPC and MIMOSA XML-based open standards to enable a proper synthesis of Operations and Maintenance (O&M) related information in an open, multi-vendor environment.

The World Wide Web Consortium (W3C) working groups target to ensure seamless Web access with all kinds of devices and worldwide standards for the benefit of Web users and content providers alike. The important activity of W3C within Synchronized Multimedia Activity is on the Synchronized Multimedia Integration Language (SMIL) for choreographing multimedia presentations where audio, video, text and graphics are combined in real time. SMIL is a W3C recommendation that has to be strongly observed by the development of the CWE solutions. The Synchronized Multimedia Working Group has been building SMIL 2.1, which defines new profiles for the mobile industry and should enable other standards bodies (*e.g.*, 3GPP, OMA) to base their multimedia messaging service related specifications on SMIL 2.1.

The Semantic Web Activity develops specifications for technologies ready for large-scale deployment and identifies infrastructure components through open source advanced development.

The main technologies of Semantic Web fit into a set of layered specifications. The current components are Resource Description Framework (RDF) Core Model, RDF Schema language and Web ontology language (OWL). Building on these core components is a standardized query language, the Simple Protocol and RDF Query Language, enabling the "joining" of decentralized collections of RDF data. These languages build on the foundation of Uniform Resource Identifier (URI), XML and XML namespaces.

A very important W3C document is Web Services Addressing, which provides transport-neutral mechanisms to address Web services and messages. Web Services Addressing 1.0 Core defines a set of abstract properties and an XML Infoset representation to reference Web services.[34]

Reference architectures. In order to enable efficient design of different systems such as CWE, automation systems, and processes, it is necessary to establish a consistent development methodology. A key tool of such a methodology is the definition of a Reference Architecture (RA) for these systems, which should provide a unified representation of essential features. The basic objectives of RA are to (Dornier 1992):

[34] This specification enables messaging systems to support message transmission through networks that include processing nodes such as endpoint managers, firewalls, and gateways in a transport-neutral manner, which is of huge importance for the complex virtualised collaboration environments. The XML Key Management Activity specifies protocols for distributing and registering public keys, suitable for use with the standard for XML signatures defined by W3C and the Internet Engineering Task Force and its companion standard for XML encryption.

- Provide a simple, unified architecture for systems, enabling traceability between the solution independent requirements and final implementations
- Achieve minimum complexity and design simplification
- Support for interfacing
- Support re-usability of modules, *etc.*

A RA may serve as a tool for the comparison of different concepts, and/or implementations, their evaluation, and an integration of modules developed within different approaches. It shall enable and efficiently support the communication between people involved in the development process. It shall serve as a means to ensure a common unified, unambiguous and widely understood terminology between people interested, from a utilization point of view, in the behavior of the system, and people, being experts in different technical domains, realizing a specific functionality of the system to achieve the required behavior and performance. To reach this objective, an RA shall be easily interpretable and applicable and, therefore, it shall represent a structure of limited complexity (Stokic *et al.* 1994).

Specifically the need for RA for CWE comes from the requirement to achieve a global agreement on how collaborative environments interoperate in order to create a critical mass of developers and users – similar to the World Wide Web. A way to specify this agreement is to define an RA for collaborative environments (Decker 2006). Therefore, RA for CWE is of a high relevance for both users and developers of CWE solutions (Correia *et al.* 2008a).

The objective is to provide seamlessly and ubiquitously a CWE to the users, thanks to a collaborative service infrastructure and end-user application-specific interfaces. For this, collaborative service interoperability is a crucial goal (Ralli 2005.

The reference architecture for CWE assumes that a high-level collaboration middleware, or e-collaboration UpperWare, is a software generic infrastructure where the generic model of collaborative infrastructure is mapped to the services; furthermore, to alleviate the tasks of the developers, the basic collaborative functions, locking, presentation control, user presence management, organization management, and communication control are gathered into a collaborative infrastructure and made available to the applications. Also a collaborative service can be built by composing or by orchestrating the (distributed) collaborative functions provided by the collaborative infrastructure.

Current middleware definition as a "glue" to patch machines and their software functions together is not appropriate to distinguish this new conceptual approach. Next to a fundamental architectural blueprint for RA it is useful to include vertical "plug-ins" as extensions for various, often changing, industrial requirements which are not an integral part of the fundamental layer (Expert Group 2006).

Most of the architectures for CWE proposed so far share three-layer models where basic technologies are at the bottom, collaborative tools stand for the middleware, and user interfaces complete the top level (see Fig. 4.2).

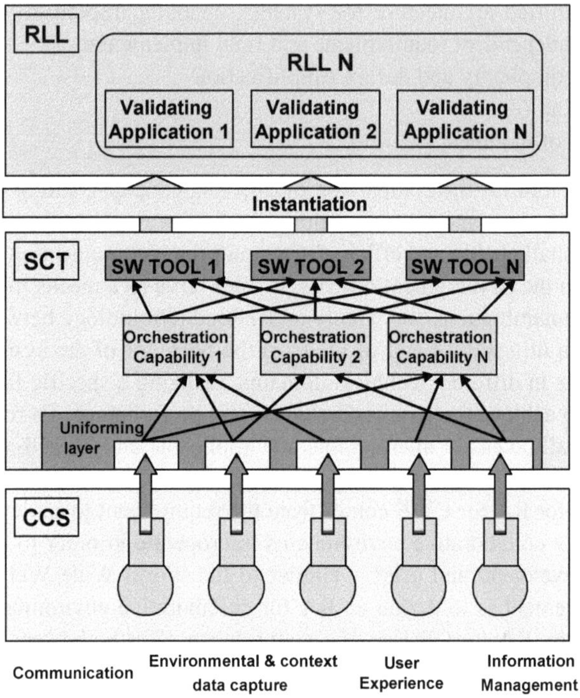

Fig. 4.2 Architecture of the C@R project and initial reference architecture for CWE (Ralli 2005)

Different capacities, functions, and features are considered in several ways, depending on each specific initial approach and therefore a general layered meta-architecture is to be worked out. Several architectures are currently available.[35]

The key elements of the CWE infrastructure are indicated at Fig. 4.3. All functions have to be offered as services to the user or application developer:

- Generic services
- Domain-specific services
- Context-specific services

[35] The one proposed by C@R project (http://www.c-rural.eu/) also served as an initial RA for CWE. The objective of the project CoSpaces (http://www.cospaces.org) is to develop collaboration models and distributed technologies to support collaborative workspaces in distributed virtual manufacturing enterprises. The development of CoSpaces architecture is based on the Open Group Architecture Framework (http://www.opengroup.org/togaf/) methodology. TOGAF is a framework for developing architectures. It provides tools to design, evaluate and build architectures for different purposes. The ECOSPACE project (http://www.ip-ecospace.org/) aims to build an infrastructure for the so-called CWE for e-Professionals, starting from such concepts as SOA, Web 2.0 and the Semantic Web.

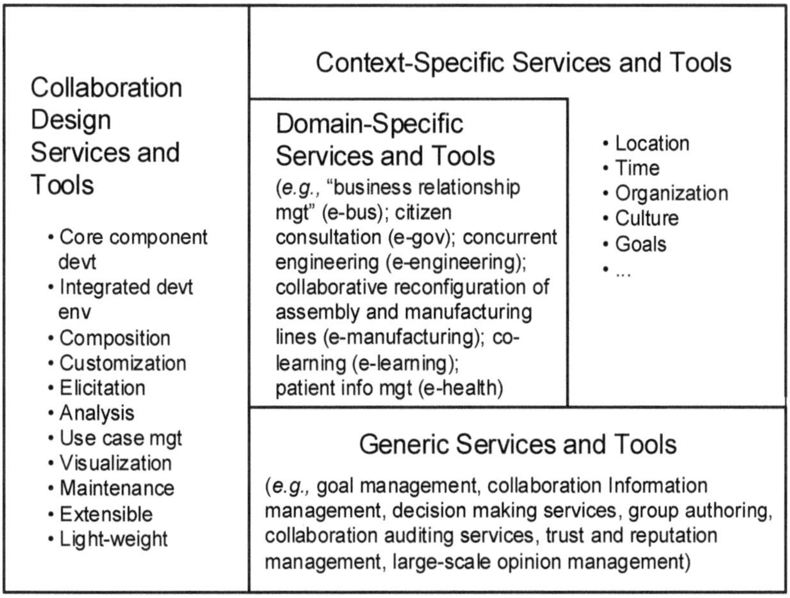

Fig. 4.3 Possible collaborative infrastructure (Expert Group 2006)

A combination of generic cross-domain services with domain-specific ones builds application-specific collaboration environments. A combination of generic and domain specific services with context-specific services builds situational specific collaborative applications. This reflects two key aspects of CWE – a key dual approach:

- Generic cross-domain aspects
- Domain, application specific aspects (Expert Group 2006)

The architectures described in Chap. 5 fit with this generic reference architecture.

4.2.5 *Security, Trust, Privacy, and Intellectual Property Rights*

Application of ICT in general and especial Collaborative Working Environments (CWE) has strong implications regarding:

- *Security*
- *Trust*
- *Privacy and Intellectual Property Right (IPR) issues*

The three aspects are quite interrelated and are going to be discussed all together in the text to follow.

Security breaches originating from the use of ICT and especially CWE have the potential to cause major losses for enterprises today, with the risks certain to increase as businesses in EE tend to increase complexity of interconnections among ICT systems and processes.[36]

It could be stated that most ICT systems today are insecure because they were not originally designed with security fully in mind. The underlying system software was designed to assure performance, size, or other aspects. Security issues have been managed with a "fail first/patch later" approach used to plug security holes. Security enabling greater trust has been added on in the form of patches or fixes (CuteLoop 2008).

In addition, the cost of evaluating and verifying system security at the highest levels has been prohibitive because the size and complexity of the software code to be evaluated was too large, making the process too slow and expensive for all but the most mission-critical systems which have been mostly funded by government for defense systems.

Intensive research on aspects of *security and trust* within EE has been going on. The challenge for businesses today has been to provide high-assurance security for distributed systems that reduces the duration, development schedule risk, and cost of designing, evaluating, verifying, and deploying secure systems (Brookson and Zumerle 2006).

Companies participating in EE need trusted technologies that provide security, but which are flexible and adaptable. Additionally, it is important to utilize standard-based solutions to ensure that technology choices have long-term market viability and support. Security must not disrupt the work within EE or cause undue burden on and complexity to existing business processes. There are three major challenges that any security framework must address:

- Authentication
- Data protection
- Access control

There is a need to deploy *trust and protection* standards as part of CWE technologies and to communicate security requirements to all EE actors to ensure that sensitive information remains protected. Traditional security is the management of risk. This involves looking at the information being protected, the threats to the system based on the environment, and the security objectives of the system. Appropriate security mechanisms and management controls are then developed and put into place to address these threats.

The planning and implementation of a security infrastructure supporting cooperation and information sharing within EE covers a wide range of processes and technologies. To be successful a number of key elements must be integrated and addressed (CuteLoop 2008):

---

[36] In the US, the National Institute of Standards has identified 30 distinct categories of threats to information infrastructures, ranging from operator errors to hacker intrusions to viruses.

- Controls should be a balance of procedural, management and technology measures. Further, the selected measures should complement one another at all levels.
- Security measures should reflect both the working practices of the organization and the controls required with the minimum possible effort and impact on the users of the system.
- Controls should meet any constraints imposed by the current technical infrastructure (including operating systems and networks) and be capable of supporting any planned migration program.
- Measures need to be defined correctly for each operating area that both complies with the organization's standards and meet local business and operational needs.
- Staff involved at all stages of the process need to be identified and equipped to meet their defined responsibilities for managing security.
- Procedures and practices need to be documented to allow the successful definition, implementation, management and long term support of the measures.

All of these areas need to be considered to provide a minimal yet balanced set of security mechanisms and controls.

For example, *Multiple Independent Levels of Security (MILS)* is an architecture that effectively addresses domain separation which is likely tat can be effectively applied (also) in EE context.[37] This architectural approach is designed to build scalable systems appropriate for the information assurance needs of both commercial and government systems (Patel *et al.* 2002). MILS uses the technique of distributed trust by placing the trusted functions and security policy throughout the system enabling the ICT system architect and security architect to work in a collaborative manner to identify realistic trade-offs in the software and ICT system design so that optimization of performance and trust can be incorporated early in the requirements and ICT system design process.[38] The MILS Architecture can be scaled from embedded system devices to complex control and transaction systems. Although initially intended for defense applications, MILS has been shown to provide a foundation for various critical systems. MILS provides a highly modular, flexible, component-based approach to distributed high-integrity systems.

As indicated above, it is of high importance to apply existing standards for addressing security within ICT systems in EE. ISO/IEC 17799, a Code of Practice

---

[37] The MILS principles were first formulated in the work of Dr. John Rushby (Rushby 1981) who published a paper in 1981 (http://www.csl.sri.com/papers/sosp81/) proposing that security is best handled by having known and limited physical interfaces between processing entities in a system.

[38] MILS applies long-established principles and techniques alongside recent advancements in microprocessors, computer security, software engineering, and formal methods. One of the tenets of MILS is to decompose a system into reusable components, each of which is simple enough to be rigorously analyzed for security properties. MILS architecture also separates security mechanisms into manageable components. Processes are isolated into partitions that comprise a collection of data objects, code and system resources.

for Information Security Management[39] is an internationally recognized set of security best practices.[40]

The control objectives and controls in ISO/IEC 17799 are intended to be implemented to meet the requirements identified by risk assessment. ISO/IEC 17799 is specifically intended as a common basis and practical guideline for developing organizational security standards and effective security management practices, and to help build confidence in inter-organizational activities.

Attention has to be given to *privacy and IPR issues* being one of the most critical aspects of collaborative work on networked industry as addressed in "Securing Intellectual Property in Collaborative Environments: a Guide" (Tang and Molas-Gallart 2004). Privacy and IPR issues are specifically critical when applying CWE for product/process development and innovation.

CWE allow for effective monitoring and transparency of the collaborative work on product/process development. However, monitoring of either data related to persons or data related to collaborative work raises several issues. The advanced technology and the increasing use of these means to "monitor" persons within, *e.g.*, collaborative product development, raise the concern that the privacy of individuals is not being respected. Privacy and property rights issues are important in manufacturing and in modern human-centric environments.[41]

4.3 Ontologies in Product/Process Innovation

Ontologies aim at capturing domain knowledge in a generic way. An ontology provides a generally (jointly) agreed understanding of a domain, which can be reused and shared across applications and groups. Such an agreed understanding of a domain – conceptualization – is necessary for every communication process. An ontology allows people to reason about sameness as well as diversity of concepts

[39]http://www.iso.org/iso/support/faqs/faqs_widely_used_standards/widely_used_standards_other/information_security.htm

[40] ISO/IEC 17799 specifies guidelines and general principles for initiating, implementing, maintaining, and improving information security management in an organization. They provide general guidance on the commonly accepted goals of information security management. ISO/IEC 17799 contains best practices of control objectives and controls in several areas of information security management: security policy, organisation of information security, asset management, human resources security, physical and environmental security, communications and operations management, access control, information systems acquisition, development and maintenance, information security incident management, and business continuity management, compliance.

[41]Projects such as CONNECT, Semantic-enabled context-sensitive privacy technologies for ambient intelligence applications (FP6-2004-IST-026464) http://www.ist-connect.eu/, which develops a privacy management platform within pervasive mobile services, coupling research on semantic technologies and intelligent agents with wireless communications and context-sensitive paradigms and multimodal (voice/graphics) interfaces to provide a secure framework, tries to ensure that privacy is a feasible and desirable component of future ICT applications.

and to derive mapping for establishing semantically correct communication channels. Hence, ontologies provide a common vocabulary of an area and define, with different levels of formality, the meaning of terms and relations between them.

Ontologies are usually organized in a classification system, and typically contain modeling principles such as classes, relations, functions, axioms, and instances.

One of the main motivations to build up ontologies is to enable sharing and re-use of knowledge and reasoning behavior across domains and tasks. Ontologies can be seen therefore as complementary reusable components to construct knowledge systems. Ontologies can be seen as a key means to assure "real" interoperability of the ICT systems.

Ontology can be a means to enable knowledge exchange among different partners of an Extended Enterprise (EE). It can even help in the translation of words between different languages. Therefore, ontologies are of key importance for modern collaborative product/process innovation (Kuczynski *et al.* 2005).

This section is dedicated to the application of ontologies for innovation processes in industry. Requirements of ontology to support innovation processes are discussed an overview of the approaches/tools for ontology building/maintenance is provided. Special attention is given to possible approaches to use ontologies in innovation processes in EE context.

4.3.1 Requirements on Ontology for Innovation

Methods for ontology building are of special relevance for innovation processes since the built ontology serve as a basis to put together and re-use innovative ideas from different actors within an EE. Current KM systems often inhibit innovative ideas within an EE, since the necessary adaptation and enhancement of the KM systems, according to the modified environment, needs often expert know-how and capacities that are not available in the enterprise. Knowledge management intrinsically involves communication and information sharing which can be strongly affected by the context in which it is viewed and interpreted. This situation gets worst when complex domains are considered, as it is the case of the industrial (and especially manufacturing) domain and complex innovation processes. The development of ontologies to unify and to put into context the different concepts and terms of the industrial domain can be very helpful to avoid misinterpretations. Therefore, application of ontologies is promising to solve basic problems of sharing knowledge within collaborative innovation processes in an extended and distributed enterprise.

The main requirements are:

- Ontologies for reuse of innovative ideas from different actors within an EE
- Uniform ontology which is applicable for different (cooperating) industrial sectors
- Enable use and enhancement/maintenance of ontology concept by company staff
- Easy and comprehensible tool to build new and to enhance/maintain existing ontologies by company staff

Innovation knowledge is often abstract and is mostly generated unstructured using unclear definitions and terms. The usage of appropriate hierarchical structures and unified terms, defined by an ontology, is promising for the efficient exchange, re-use, further elaboration and exploitation of innovations (Meier *et al.* 2001; Sorli *et al.* 2002a). However, the challenging problem is how to use such a structured approach, which obviously allows effective collaborative work within a network, but which does restrict collaborative creativity (Nemiro 2004).

The collaborative exploitation of the product and process related innovation knowledge within an EE or a network of enterprises and specifically within collaboration networks established by SME is of the highest significance. Since innovations can be specified multifaceted and unstructured and are imprecise (Meier *et al.* 2001), the objective to store, re-use, and develop innovations with IT means can be considered even nowadays as very challenging. IT systems have to regard the adverse requirements of innovation knowledge definition in order to assure exploitation of innovations within industrial companies in an efficient way.

Additional requirements on ontologies and tools for ontology management to support innovation process in an EE context are:

- Support structuring of unclear terms
- Support multifaceted definition of terms
- Do not restrict creativity.

4.3.2 Methods and Tools for Ontology Building/Maintenance

Ontologies attract close attention of the RTD community. Ontologies are considered to be a powerful solution to the problem of efficiently storing and retrieving knowledge, currently increasingly investigated in the scope of research on semantic Web, promoted by many researchers and also recognized by industry (Gómez-Pérez *et al.* 2004). A number of ontologies and tools to setup/maintain ontologies have been developed or are currently under development. Numerous tools and applications of semantic Web technologies are already available and the number is growing fast. However, their application in practice is still not widespread and application-oriented methods for product and process innovation domain are still under development.

ICT tool support is important both for the ontology development process (ontology building, annotation, merge, *etc.*) and for the ontology usage in applications such as electronic commerce, the Semantic Web, KM, specifically for KM within innovation process in EE context. When one is about to build an ontology, several basic questions arise related to the tools to be used (Soares *et al.* 2005):

- Which kinds of tools are the most convenient for building an ontology?
- Which tools give better support to the ontology development process?
- Does the tool have an inference engine?
- How can applications interoperate with ontology tools?
- How can the developed ontologies be used in real applications?
- How can one reuse other existing ontologies in the same domain?
- How can one merge two similar ontologies built for the same domain?
- How can one evaluate the quality of the developed ontology or other existing ontologies to be reused?
- What is the stability and maturity of an ontology tool?

The methods and tools for ontology building/management can be structured in several key groups. In the text to follow, a brief overview of these groups is provided, including some typical examples of existing methods/tools[42] (Soares *et al.* 2005):

Ontology development methods and tools. This group includes tools, environments, and suites that can be used for building a new ontology from scratch or reusing existing ontologies. Apart from the common edition and browsing functionality, these tools usually include ontology documentation, ontology exportation and importation from different formats, graphical views of the ontologies built, ontology libraries, attached inference engines, *etc.* There are many available tools on the market, and some of them are:

- *Apollo*[43] is a user friendly ontology development application. The design was motivated by the developers' experiences working with industrial partners who wished to use knowledge modeling techniques, but required an easy to use and understand syntax and environment.[44]

---

[42] As already indicated, the authors by no means intend to recommend the tools for specific applications or use. The tools are listed only as examples to explain better different possibilities to support ontology building and use.

[43] http://apollo.open.ac.uk/docs/user-guide.ps.gz

[44] Apollo supports all the basic primitives of knowledge modelling: ontologies, classes, instances, functions, and relations. Full consistency checking is done while editing, for example, detecting the use of undefined classes.

- LinKFactory®[45] is a formal ontology management system, designed to build and manage very large and complex language-independent formal ontologies (Jackson and Ceusters 2002).
- OntoEdit®[46] is an Ontology Engineering Environment supporting the development and maintenance of ontologies by using graphical means. OntoEdit is built on top of a powerful internal ontology model. This paradigm supports representation-language neutral modeling for concepts, relations and axioms.
- *Ontolingua Server* is a set of tools and services that support the building of shared ontologies between distributed groups. The ontology server architecture provides access to a library of ontologies, translators to various programming languages, and an editor to create and browse ontologies.
- *KnoME* is a large suite of tools for the collaborative development of ontologies in the GRAIL concept modeling language (Rector *et al.* 97). These tools are freely and openly available as OpenKnoME (Rogers *et al.* 2001).
- *Protégé-2000* is one of the most popular ontology development tools. It has thousands of users all over the world who use the system for projects ranging from modeling cancer-protocol guidelines to modeling nuclear-power stations. Protégé-2000 provides a graphical and interactive ontology-design and knowledge-base–development environment. It effectively helps knowledge engineers and domain experts to perform knowledge-management tasks (Noy and Musen 2001).
- *SymOntoX* (Symbolic Ontology Manager XML savvy) is a software prototype for the management of domain ontologies. In SymOntoX, domain concepts and relations are modeled according to OPAL (Object, Process, and Actor modeling Language), a methodology for ontology representation developed for tourism domain (Missikoff *et al.* 2002).
- *WebODE* (Arpírez *et al.* 2001) is an ontological engineering workbench that provides varied ontology related services and covers and gives support to most of the activities involved in the ontology development process and in the ontology usage.[47]

Ontology merge and integration tools. These tools solve the problem of merging or integrating different ontologies on the same domain. This need appears when two companies or organizations are merged together or when it is necessary to ob-

[45] LinKFactory® is a registered trademark of Language and Computing, Inc., 6701 Democracy Blvd., Suite 300, Bethesda, MD 20817 U.S.A. http://landcglobal.com

[46] OntoEdit is a registered trademark of ontoprise GmbH, Amalienbadstr. 36, (Raumfabrik 29), 76227 Karlsruhe, Germany http://www.ontoprise.com

[47] It is built on an application server basis, which provides high extensibility and usability by allowing the easy addition of new services and the use of existing services. WebODE ontologies are represented using a very expressive knowledge model, based on the reference set of intermediate representations of the so-called METHONTOLOGY methodology (Fernández-López *et al.* 1999).

tain better quality ontology from other existing ontologies in the same domain. Ontology merging in design time is very important, since building of an EE or company networks can lead to a merging of their ontologies. It can even be necessary to merge several ontologies to have another with better quality. Ontology merging in run-time can be also crucial. In fact, ontologies have become increasingly common on the World Wide Web where they provide semantics for annotations in Web pages (Noy and Musen 2001). Moreover, the heterogeneity of information on the Web can provoke one to deal with different ontologies in similar domains. Such diversity must coexist with the interaction between systems. An option to make compatible the diversity and the generality is to establish mappings between ontologies, and to merge them in run-time (Mena and Illarramendi 2001).

Some ontology merge and integration tools are briefly described below:

- *Chimaera* is a merging and diagnostic Web-based browser ontology environment. Its design and implementation is based on previous experience in developing other user interfaces for knowledge applications such as the Ontolingua ontology development environment (Farquhar *et al.* 1997), collaborative environment for building ontologies for FindUR (McGuinness 1998), and other systems.
- *FCA-Merge* is a method for merging ontologies which follows a bottom-up approach offering a global structural description of the merging process. For the source ontologies, it extracts instances from a given set of domain-specific text documents by applying natural language processing techniques.[48]
- *PROMPT* (Noy and Musen 2000) is a tool for semi-automatic guided ontology merging. It is a plug-in for Protégé-2000 (see above). The user makes many of the decisions and PROMPT performs additional actions automatically based on the user's choices and creates a new set of suggestions and identifies additional conflicts among the input ontologies.

Ontology evaluation methods and tools. These support tools ensure that both ontologies and their related technologies have a given level of quality. Quality assurance is important to avoid problems in the integration of ontologies and ontology-based technology in industrial applications.

Because of the size of ontologies, their complexity, their formal underpinnings, and the necessity to come towards a shared understanding within a group of people, ontologies are still far from being a commodity. Developing and deploying large-scale ontology solutions typically involves several separate tasks and requires the application of multiple tools. Therefore pragmatic issues such as interoperability are key requirements if industry is to be encouraged to take up ontology technologies rapidly.

[48] Based on the extracted instances, mathematically founded techniques taken from *Formal Concept Analysis* (Wille 1982; Gartner and Wille 1999) are applied to derive a lattice of concepts as a structural result of FCA-Merge. The produced result is explored and transformed to the merged ontology by the ontology engineer.

The large visibility of the semantic Web, its tools and applications already attract industry. In particular, as they move from academic institutions into commercial environments they have to fulfill stronger requirements and in some cases new requirements (*e.g.,* concerning scalability and multi-user access). Different tools from different sources need to interoperate and therefore are typically no longer standalone solutions but integrated into frameworks. These frameworks must be open to other commercial environments and provide connectors and interfaces to industrial standards. Larger applications also need larger ontologies and therefore require substantially better performance and scalability.

A systematic evaluation of ontologies and related technologies might lead to a consistent level of quality and thus better acceptance by industry. This effort leads to standardized benchmarks and certifications. Examples of ontology evaluation methods and tools are:

- *OntoAnalyser* and *OntoGenerator,* are realized as plugins for OntoEdit which is a graphically oriented ontology engineering environment[49]
- *OntoClean in WebODE:* series of ontology evaluation criteria based on philosophical notions as: rigidity, identity, unity, dependence, *etc.* have been elaborated (Guarino 2002) and used in OntoClean, the ontology evaluation methodology to clean tangled ontologies
- *ONE-T* (Ontology Evaluation Tool) is a Java Web-based application that allows verifying ontologies stored and available in any Ontolingua server

Ontology-based annotation methods and tools. These tools have been designed to allow users to insert and maintain (semi)automatically ontology-based markups in Web pages. Most of these tools have appeared along with the emergence of the semantic Web. Most of them are integrated in an ontology development environment.

Ontology storage and querying methods and tools. The tools have been created to make using and querying ontologies easy. Due to the wide acceptance and use of the Web as a platform for communicating knowledge, new languages for querying ontologies have appeared in this context.

Ontology learning methods and tools. These tools are used (semi)-automatically to derive ontologies from natural language texts.

---

[49] The underlying plugin framework allows for flexible extensions of OntoEdit's core functionalities (also for third parties). This allows tight integration of the evaluation of ontologies that are (i) created with OntoEdit or (ii) imported from other tools like Protégé and WebODE into the development process. Each of the two plugins addresses different aspects of the evaluation criteria. OntoAnalyser focuses on evaluation of ontology properties, in particular language conformity and consistency. OntoGenerator focuses on evaluation of ontology-based tools, in particular performance and scalability.

4.3.3 Ontologies for Innovation in Extended Enterprise

It is clear that ontologies are important tools to support innovation in Extended Enterprise (EE) context. The problem is how to set-up and manage ontologies for this purpose.

The advantages of using a single ontology at the heterogeneous, federal IT systems of the collaboration partners within an EE are obvious, but such a concept has its drawbacks: every partner has to base his information system solely on the structure defined by the ontology which will hinder him in using existing information repositories, refining the structures according to his needs, and adding further partners to his network (Wache *et al.* 2001). An alternative approach suggests allowing every partner to operate his information systems based on his own ontology.

Another concept has been proposed to avoid the above – mentioned problems by using a single ontology: the concept is to use a *hybrid ontology approach* with a "shared vocabulary" (Wache *et al.* 2001; Visser *et al.* 2002; Uschold 2000) as depicted in Fig. 4.4. This concept defines a unified terminology (or "vocabulary") which is a ground for the exchange of information supplied with semantics. It is also possible to use instead of the "shared vocabulary" a "shared ontology" also called "global ontology" (Uschold 2000), a standardization of information's structuring, meaning, relationships, and constraints to express knowledge efficiently.

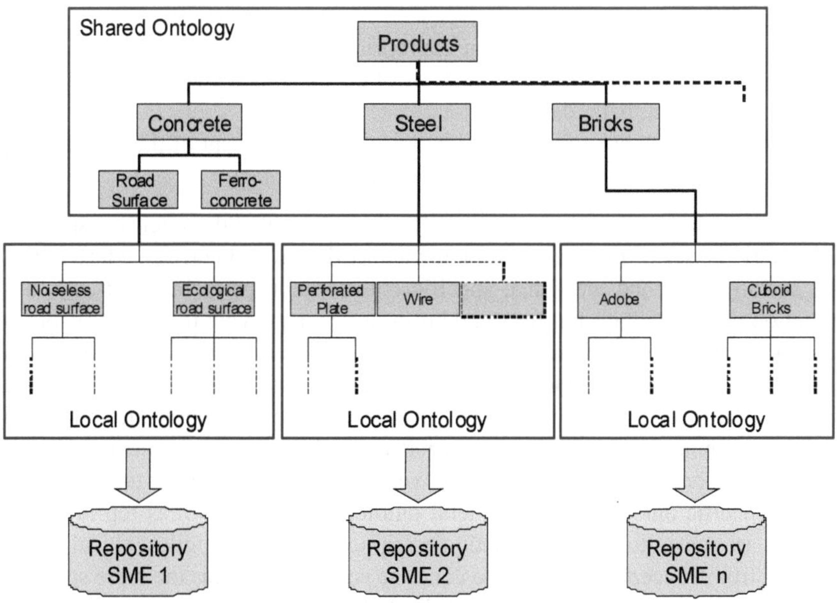

Fig. 4.4 Concept of the hybrid ontology approach

The interchange of the data among the partners is organized by the so - called mediators, programs which realize the mapping between heterogeneous information systems. The main handicap of this mediator is the effort to establish the mapping rules which are very extensive and cumbersome to define. Approaches exist to realize this to an extent autonomously (learning or systematically analyzing), *e.g.*, such an approach is proposed by Wache (2003), but they cannot guarantee a full seamless information exchange.

On the other hand, the local repositories could be adjusted to comply with the global ontology. This concept's main drawback is the amount of effort for its realization, because adjustments to the existing local repository structures are cumbersome and may be erroneous, and building up a compliant local repository anew with import of existing data can be costly. However, this concept seems to be most promising for a practical application in the heterogeneous collaboration networks of the companies or EE. It supports effective sharing of knowledge by the agreement of the shared ontology and also allows using existing data repositories which were built up individually.

To apply such a concept at an EE or collaboration network with already existing knowledge repositories at different nodes, an approach has been developed to adjust gradually existing local ontologies at the local repositories (at different nodes of the collaboration network) so that they comply with the hybrid ontology approach. This approach comprises a methodology and IT architecture to solve the problem of adaptation and re-use of existing local ontologies for the hybrid ontology concept. The methods and tools provide (Kuczynski *et al.* 2005):

- A support in the definition of an appropriate global ontology, addressing innovation knowledge, based on analysis of existing local ontologies of the networks or EE partners' databases
- Set-up of the global ontology and maintenance/adjustment of the partners' local ontologies, as well as of the global (shared) ontology

The global ontology must be designed and implemented thoroughly respecting the specific issues of the innovation knowledge, design and corresponding industrial sector(s), and the individual specific collaboration partners' needs, *i.e.*, information needed in their business and business segments with their specific products and activities. But it has to be general and flexible enough to cope with requirements from new partners and changes in business types. In short, the global ontology must be flexible and precise enough to store efficiently the individual participants' innovation knowledge at persistently changing requirements.

In general, ontologies may range from very general ontologies up to application domain specific ontologies. The general ontologies, also known as top-level ontologies, are meant to provide holistic descriptions, spanning philosophical (starting, *e.g.*, from concept "thing") up to concrete concepts whereas the more specific, often called "domain ontologies," are dedicated for different domains, *e.g.*, for:

- Enterprise communication, *e.g.*, Uschold's *Enterprise Ontology* (Uschold *et al.* 1998)
- Knowledge management
- Other domains, *e.g.*, medical, pharmaceutical, engineering, law, geography, *etc.*; see (Gómez-Pérez *et al.* 2004) for an overview

For innovation process support in (manufacturing) industry the general Enterprise Ontology (Uschold *et al.* 1998) may be chosen consisting of a comprehensive collection of terms and definitions relevant to business enterprises, providing sufficient flexibility and generality to define business entities and activities. For the product aspects the ontology relevant for the specific EE or network has to be applied. Based on the Enterprise Ontology and product related ontology(ies), any of the many different ontology modeling methodologies existing may be re-used (Gómez-Pérez *et al.* 2004) to redefine and elaborate the complete global ontology. The ontology modeling methodology of Uschold can be applied, leading to the following general steps:

- *Activity "Ontology capture"* (Gómez-Pérez et al. 2004):
  - Analysis of companies' business domains, definition of common practices important for most of the partners (step "To identify the purpose and the scope")
  - Analysis of innovation knowledge terms to describe characteristics of product or process improvements and definition of global terms to define innovations (step "To build the ontology")
  - Analysis of terms used in companies' business and definition of central, basic terms (terminology), common for the most companies (step "To build the ontology")
  - Analysis of product definitions and standards and the terminologies for specifications and classifications (step "To build the ontology")

- *Activity "Integrating existing ontologies":*
  - Comparing the collected terms and relations with the enterprise ontology and product ontology concepts and refinement of concepts for the global ontology (step "To build the ontology")

For each company, an individual set of terms and their usage will be analyzed. The fourth of the five sub – steps above is the most important, trying to find terms in common and similarities to other terms in order to define an initial set of terms useful and common to all companies to describe innovations efficiently.

Since descriptions of innovations are often unstructured, ambiguous and fuzzy (as far as no definition structures are assumed), the fourth of the five sub – steps above has to be emphasized to focus the modeling efforts on the analysis of business processes and products in order to acquire more detailed ontology models:

- Analysis of processes including technology used, production sequences, production units and production additives
- Analysis of products including design defaults and constraints (technological, functional, appearance requirements, *etc.*), failures and weaknesses (reports about production errors, quality problems, customers' feedback, *etc.*)
- Analysis of products' exploitation including models about sale markets, market needs, expected price, expected benefits, impact on market and expected relevance for new patents (similarity to former patents)
- Analysis of innovations including efforts (estimation of complexity, requirements, comparison to development of former innovations), efforts related to other partners (communications and arrangements needed), feasibility (degree of maturity), modifications and alternatives (relations to former or similar innovations), synergy from other innovations, change management including change impact, change frequency, *etc.*

The resulting set of common terms together with their relations and the selected top-level ontology will establish the global ontology, which shall be compliant to the collaboration partners' business domains as well as providing specific definitions to capture innovation knowledge efficiently.

The presented methodology and the ICT system effectively support the collaborative exploitation of innovation knowledge by applying the hybrid ontology concept within a network of SME. The results have been specifically examined within the SME network in construction industry,[50] but the approach is applicable to different SME networks. The methods and the system architecture allow for en efficient introduction of the global ontology related to innovation knowledge in a collaboration network of SME with heterogeneous data repositories.

The advantages of this approach to harmonize continuously the distributed innovation knowledge bases (local databases) in order to achieve an accelerated exchange of innovation knowledge within a collaboration network or an EE in comparison to current practices in industry are obvious: the exchange of knowledge on innovation by the structured approach, based on a global ontology, is assumed to be more efficient than "informal" approaches in current practice.

---

[50] See project Know-Construct: http://www.know-construct.com/index.htm

Chapter 5
ICT Tools for Collaborative Product/Process Design and Innovation Process

Abstract The modern approaches for product/process development and innovation in industry require involvement of many different teams in extended enterprise (EE) context. The increasing trend of globalized industrial environments and a radical increase in number of product variants in modern industry (*e.g.*, build-to-order) requires new forms of collaboration among teams involved along product/process innovation life cycle (*e.g.*, design, planning, production scheduling, manufacturing, after sales services, *etc.*), as well as seamless knowledge and experience sharing among these teams, often distributed geographically and in time. The chapter focuses upon information and communication technologies (ICT) support for collaborative innovation. It discusses how so-called Collaborative Working Environments (CWE) support collaboration among teams in industry enabling more effective innovation processes. The chapter addresses design/improvement/reengineering of product/process and innovation processes in general. First, general aspects of collaborative work in industry are examined, aiming to identify common collaboration features that can be supported by CWE. Then an ICT platform for collaborative product/process design is discussed. The approach is extended to collaborative innovation process in EE context. Several CWE solutions supporting collaborative product/process development and innovation process are described. Specific aspects related to the ICT support for collaborative innovation in small and medium sized enterprises (SME) driven Extended Enterprises are discussed.

5.1 Collaborative Work in Industry

As indicated in previous chapters, collaborative work is crucial for modern industry active in the global market. As explained in Chap. 3, one of the key elements of the new model for product/process innovation, and specifically of systemic innovation, is collaborative work. The collaborative work in industry includes:

- Collaborative work within organized teams in industry in *Extended Enterprise (EE)* context
- Collaboration between teams in industry and *ad hoc* groups and wider communities, *e.g.,* RTD, consumers, *etc.*

As indicated in Chap. 4, modern ICT, *i.e.,* Collaborative Working Environments (CWE), may effectively support such collaborative work. The analysis of industrial requirements clearly indicates the need for different information and communication technologies (ICT) services supporting the collaborative work among different groups and satisfying the following (basic) requirements:

- Support different collaboration patterns with special emphasis on temporal aspects (*i.e.,* synchronous, asynchronous as well as multi-synchronous patterns), since the collaborative work in industry often requires a combination of different temporal patterns
- Support distributed work in EE environment, which includes:
  - Identification of appropriate expertise
  - Team forming
  - Checking availability *etc.*

- Support effective (on-line) provision of information/knowledge on product/processes and on actors involved, needed for collaborative work
- Support effective knowledge management (knowledge acquisition, discovery, and presentation) including decision making
- Support dynamic changes in collaborative work conditions

It is obvious that the different collaborative activities – *i.e.,* the corresponding SW services to support these activities – have common "collaborative" actions which may be supported by a common approach.

Since the SW services have to be dynamically updated and have to be integrated with other collaborative services, there is a clear need to provide a platform which will allow effective generation/update of such services.

This section describes general aspects of collaborative work in industry and identifies common collaboration features that can be supported by CWE. Various ICT solutions supporting collaborative work and specifically collaborative innovation in industry are discussed.

5.1.1 Collaboration Patterns in Industry

Different patterns of collaboration in industry are identified. These patterns comprise temporal and spatial characterizations of the collaborative effort, as well as rules that define its boundaries and development. In order to support the different identified patterns, *ICT collaborative platforms* must provide a support for each situation, *e.g.*, by allowing users to choose the most appropriate pattern for each situation (InAmI 2006).

The main needs regarding collaborative work and respective collaborative pattern characteristics in industry include:

- Provision and selection of adequate communication tools. These requirements relate to temporal and spatial aspects of the collaboration pattern.
- Support for:
  - Collaborative start-up – contextualization
  - Identification of collaboration pattern
  - Resource discovery
  - Team definition
  - Call for collaboration
  - Collaboration start

  These requirements indicate the necessity for the definition of generic collaborative patterns with their characteristics (see the text to follow), and additionally the actors that will have the roles of collaborators in the collaborative effort.
- Tracking of actions, decisions and events: these requirements indicate the need for work rules and constraints to regulate collaboration.
- Possibility of "real-time" or asynchronous communication, of using Web services and different display devices and of modularity: requirements upon temporal and spatial characteristics.

5.1.2 Collaboration Pattern Specification

Collaboration patterns can be seen as the outline of interactions among workgroup members (Biuk-Aghai 2004). These interactions provide interdependence between collaborators and, in turn, can be seen as the necessity to share information and objectives, to divide labor, and to think explicitly in conjunction (Campos *et al.* 2006). Collaboration pattern comprises:

- Temporal aspects
- Spatial aspects

- Rules that structure the interactions among collaborating actors and the information they deal with (Wasson 2000)

Temporal aspects. Considering the *temporal* structure of the collaboration process, several types are indicated in Table 5.1 (Miao and Haake 1998; Molli *et al.* 2002).

Table 5.1 Temporal aspects of collaborative process

Synchronous processes	Data is shared by team members in the same period of time. Modifications made by one member are immediately observed by other active team members
Asynchronous processes	Each user works on the data separately. Activity parts assigned to different team members are achieved at distinct times
Multi-synchronous processes	Modifications occur in parallel; the process comprises divergence and convergence[1] cycles. These cycles occur in loops until objectives are achieved

Spatial aspects. In terms of *spatial aspects*, collaborating individuals can be:

- *Locally* concentrated, where team-members have the possibility of face-to-face contact
- *Distributed* throughout a virtual working environment, with team-members in different geographical locations and working environments (Sarma 2005)

In the latter case, collaboration lacks richness in terms of awareness and informal communications. Studies indicate that the frequency of communication between co-workers decreases with distance (Kraut *et al.* 1990). Setting up collaborative interactions for distributed work-teams implicates therefore a more complex coordination effort and also additional means to overcome the restricted flow of information and increase of social awareness (Herbsleb and Mockus 2003). Adequate tools must be available to enrich communication and information-exchange (*e.g.*, including file-sharing or chat capabilities).

"The Future Workspace – Perspectives on Mobile and Collaborative Working" (Schaffers *et al.* 2006) presents the concept of *mobility* as a key characteristic in modern working environments. Distribution inside the organization is presented in terms of "static" and "mobile" actors, instead of being classified based on geographical location. Tasks are described as:

- Context-independent (execution of tasks is independent of time and place)
- *Context-aware* (optimizing the execution by considering location information and personal profiles of users)

[1] "During divergence phases, each participant works in insulation. During convergence phases, participants synchronize their different copies to re-establish a common view of the data" (Molli *et al.* 2002).

- Context-switching (bearing different time and space context settings, including unpredictable places and times)

This leads to a categorization of workplace support including *predictability* of time and place dimensions. The concept of mobility becomes implicit in this classification. Distributed collaborative environments must be able to provide different tools, according to the context (*e.g.*, video-conferencing may be useful for synchronous distributed situations, while e-mail can be more appropriate for distributed, asynchronous tasks).

Rules: In collaboration processes, *rules* define the boundaries of the collaboration process. According to Mezura-Godoy and Talbot (2001) they can be classified as presented in Table 5.2.

Table 5.2 Classification of rules for collaboration processes

Work rules	Defined by participants; can be negotiable, therefore removed, updated or replaced during activity (*e.g.*, defining a leader of the process, deadlines for sub-tasks)
Norms	Each group members is expected to respect these; usually not explicit – usually known for all group members (*e.g.*, all participants should be addressed formally, every user will share the relevant information he/she owns)
Constraints	Not negotiable; usually established by external situations or by technical aspects (*e.g.*, imposed deadline for collaboration)

According to the collaborative scenarios detected in industry and the definition of collaborative pattern provided above, generic collaboration patterns relevant for collaborative innovation in industry are defined by the characteristics presented in Table 5.3 (Campos *et al.* 2006).

Table 5.3 Collaboration pattern (Campos *et al.* 2006)

Characteristic	Possible values	Description
Temporal	Synchronous Asynchronous – predictable Asynchronous – unpredictable Multi-synchronous Aware of time	Defines the temporal type of the pattern
Spatial	Local Distributed – predictable Distributed – unpredictable Aware of place	Defines the spatial type of the pattern
Rules	(Non-specific)	Each rule can be a work rule, a norm or a constraint

5.1.3 Generic Collaboration Pattern and Use Cases

The objective of this section is to present generic use cases that describe the interaction of human actors with an ICT platform supporting collaborative work and that are valid for the innovation processes in industry.

For the definition of these use cases, the collaboration in industry was analyzed, identifying common issues and divergences in order to model the collaboration process in industry, *i.e.,* describing the activities common to all collaboration processes and independent of a specific application or situation. This work leads to the definition of a common use case describing primarily the initiation of a collaboration process allowing a definition of common actions within different use cases which are supported by common ICT services within CWE (Campos *et al.* 2007). These services may support *Management of Social Interactions* (MSI) during the collaborative work.

Flexible and customizable collaboration platforms, not limited to specific application services, are needed. These application services often need to be dynamically updated due to changes in rules, constrains, environments, *etc.* This flexibility is fulfilled by the ability of easily updating and creating new application services. Such an approach allows users to develop and enhance their own collaborative environments thus on long term blurring the boundary between designing and using and enabling rapid context switching (Expert Group 2006).[2]

In the text to follow, two generic use cases are presented, serving as baseline for ICT platform to support collaborative work in industry and describing (InAmI 2006):

1. Collaborative process – specifically initial phase of collaborative work
2. Creation of SW service to support collaboration activities on specific applications (*e.g.,* production/process design)

Starting the collaboration process. The collaboration process starts with the identification of a situation that requires collaboration to meet a specific goal. This action is performed by an individual (or a group) that identifies the needs of the collaboration process and initiates the process in the collaborative platform.

This situation can be new or the repetition of a past situation. In both cases, the collaboration process is started up with the input by the service-starting actor of the basic data needed to start collaboration, such as:

- Topic or context
- Goal
- Effort

---

[2] However, specific conditions within industry regarding security and IPR issues have to be taken into account, which in turn may have strong influence on definition of SW services. It is likely that any enhancement of SW services in industry will ask for previous agreement of different (business) partners involved.

- Collaboration pattern for the collaboration process
- Budget
- *etc.*

From this start-up request, the platform discovers the collaboration resources, *i.e.,* expertise for a specific topic among the group of registered and general actors, first finding an optimal team composition for the specific situation and then checking their availability (present and future) and collaboration costs (direct and indirect costs of a specific actor).

After the team is composed, a collaboration call is launched to the actors and these actors confirm their availability/expertise for the subject and time frame.

According to the goal and urgency of the collaboration process, different tools and collaboration patterns are necessary for the process. The system provides product and process knowledge to support the collaboration process with all relevant information. This information can be requested directly by an actor or provided automatically by the platform in the context of the collaboration subject.

The collaboration process is supported by a set of standard communication services such as e-mail, phone, SMS, video, *etc.*

In every step of the collaboration process, the platform registers all contributions in order to trace the collaboration process and increase the knowledge of the system about the actors (*i.e.,* typical availability temporal frames).

This use case is illustrated in Fig. 5.1.

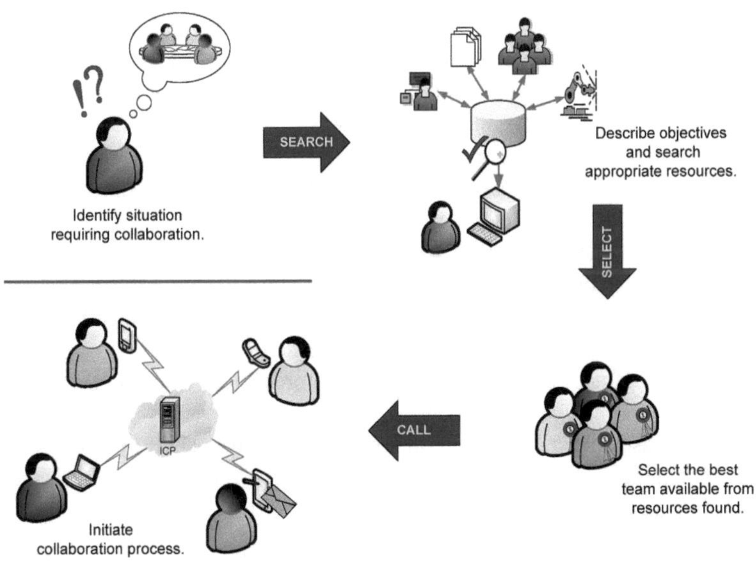

Fig. 5.1 General use case to start the collaboration process (InAmI 2006)

Creating a new collaborative application specific service. An application service is a means to access system functionalities in a prescribed way in order to save users' time and provide "guidance" for best usage of the platform. The platforms have to comprehend different application services. The unconstrained (or unguided) access to the platform disperses the common user and most of the time is spent in searching and selecting the best functionality for each specific purpose.

In the context of application services for collaborative work in industry, the creation of a new specific application service is about prearranging:

- How the collaboration will be established among actors
- Which available functionality will be used
- What type of knowledge will be provided by the system
- What type of filtering is suitable to collect relevant information from sensors
- What semantic-based knowledge management tools will be made available to process this knowledge
- How this knowledge will be presented to each type of actor
- How the conclusion of the collaborative process will be reached

Therefore, the creation of a new application service is started by a user that has identified the need for the application service. The functionality of creating an application service has to be restricted to a group of users with high access rights to the platform and might require programming skills according to the service complexity.

After inserting the basic data of the application service (title, purpose, rules, *etc.*), the user creating the application service should identify and input the best collaboration patterns to achieve the specific goal of the service, this means, to configure the temporal and spatial characteristics of the collaboration process. Another important input is the target group for the SW service, for instance, if it is directed to management, shop-floor workers, *etc.* This first configuration step is completed by the selection of the adequate functionality and knowledge management (KM) tools needed to fulfill the defined application service.

The following configuration step is to identify the knowledge required (input) and generated (output) by the application service. The relevant information on product/processes for the service purpose should also be identified together with the adequate filtering tools to contextualize this information.

The user should define which knowledge will be presented for each user group and how.

These two use cases describe in general terms possible ICT support to collaborative work in industry, and specifically to collaborative product/process design and collaborative innovation process.

5.2 ICT Platform for Collaborative Product/Process Design

This section is dedicated to ICT support for collaborative product/process design in industry. It describes how the above general considerations on collaborative processes in industry and ICT support of such processes can be applied to collaborative product/process design.

As indicated in previous chapters, collaborative product/process design is crucial for modern industry acting at the global market, *i.e.,* it is one of the crucial approaches for a new product/process design model as the generic one that has been described in Chap. 3. The increasing complexity of design processes in modern manufacturing in global economy, involving actors/teams in geographically distributed locations, has to be approached by innovative means. The problem of effective involvement of *different teams within EE* (such as product design and shop-floor teams from both system provider and components manufacturers, distributed over the globe) and *customers* in product/process design has been addressed by a number of methods and ICT tools. The modern ICT-based Collaborative Working Environment (CWE) and knowledge management (KM) technology offer possibilities to cost-effectively provide means for collaborative design process in global settings by supporting work in the "virtual world" (*i.e.,* over the Internet). However, virtual collaboration of different teams and customers is often difficult to establish, especially in sectors with complex product families asking for efficient customization/individualization of product variants and effective reconfiguration of processes to produce customized solutions.

Building of SW services to support collaborative design using CWE and KM technology requirein general, high investments and ICT specialists. Such services often have to be adapted/reconfigured to meet specific (evolving) requirements concerning specific *Extended Enterprise (EE)* or collaborative network of enterprises.

The general requirements regarding collaborative work in industry are defined in Sect. 5.1. The following are the specific requirements on ICT support for collaborative product/process design:

- The tools/services have to allow for easy *adaptation/configuration* to specific collaboration needs and *business models* within different product design processes and partnerships
- The virtual collaborative systems must support *different collaboration patterns and cultural diversity* among teams around the world, including, *e.g.,* time zone problems, Intellectual Property Rights (IPR) issues, *etc.*
- The full *traceability of design decisions*, being the prerequisite for an effective collaborative work in design of complex product families, has to be provided
- Mechanisms to support *dynamic changes* in design XE "design:collaborative"} problem definition as the collaborating actors move forward in the collaborative design process (dynamic change management), are needed

- The *selection* of tools for specific design stages and specific *EE* and collaborative situation/context is critical
- Effective (on-line) provision of knowledge on product/processes and on actors involved is a critical aspect on terms of the platform use, *i.e.,* an effective KM is needed
- Dynamic changes in collaborative work conditions during product development process have to be supported
- Collaborative decision making in EE needs to be supported

This is especially a challenge for *small and medium sized enterprises (SME)*: a cost-effective design of products with geographically distributed partners (*e.g.,* within *SME-driven EE*) and for geographically distributed customers (*e.g.,* system providers) is a key new challenge for globally acting SME (see Sect. 5.4). The specific complexity represents a *multi-threaded character of EE*, *i.e.,* the fact that each company is likely to be simultaneously involved in several EE organizations asking for tools to adapt their design environments easily.

ICT has to provide a virtual environment which is appropriate for *collaborative work* on *design in international settings*. In order to reach this target, there is a need for *platforms* based on SW collaborative services supporting design processes where different teams and customers from geographically distributed locations have to be involved.

As explained in Chap. 4, such platforms most often follow *Service Oriented Architecture (SOA)* principles, enabling *collaboration among all stakeholders* throughout the EE (including customers) or collaborative networks of companies, along the product/process life cycle, from the conceptual design up to manufacturing and utilization (feedback from customers).

5.2.1 ICT Platform Architecture

The ICT platforms allow for an easy generation/configuration of different collaborative application specific (SW) services for collaborative product/process design in international settings along the product/process development life cycle (*e.g.,* service for collaborative conceptual product design, service for collaborative manufacturability investigations, *etc.* – see the text to follow), adapted to the specific needs of different EE. Such platforms may combine (see Fig. 5.2):

- ICT tools supporting different phases of product/process design
- SW services to support Management of Social Interactions (MSI) within collaborative work
- SW services for knowledge sharing within collaborative work
- SW services to support design process (for "on-line" context-sensitive selection of the most appropriate design tools for specific design stages and specific collaborative situations)

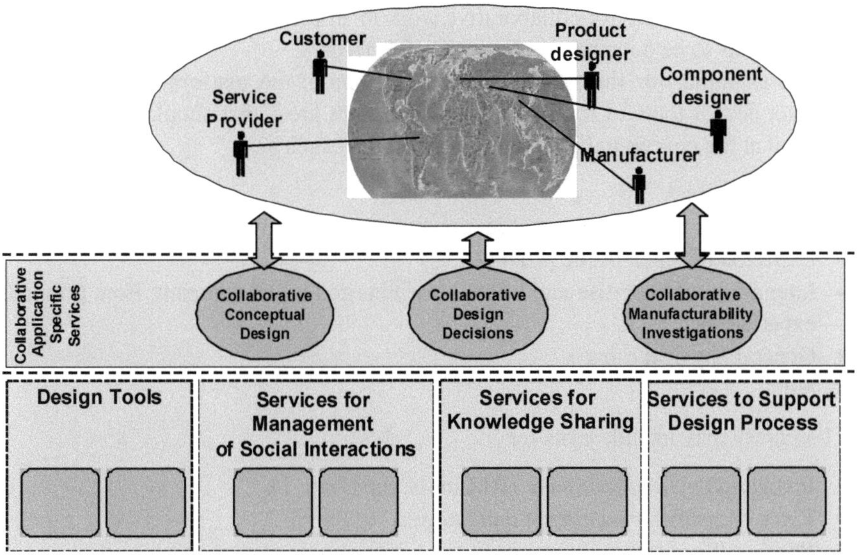

Fig. 5.2 Platform for collaborative product/process design

In addition, service engineering tools for effective generation/configuration of SW services for collaborative product design by non-IT experts fitting specific EE and product/process development needs are often parts of the platforms. Please note that the architecture in Fig. 5.2 represents an instantiation of the generic architecture shown in Fig. 4.1.

The platforms have to be open for different design frameworks, *i.e.,* they have to support different collaborative design approaches by combining different existing (classical) design tools with MSI services and KM tools/services for knowledge sharing, including services/tools for context-sensitive selection of the most appropriate design tools. The rationale behind such an approach is that companies often cannot afford to switch to use completely new collaborative design tools in a too short period, but a stepwise transition from classical design approaches to collaborative design approaches is needed. Therefore, companies need to continue to use their existing classical tools for design, which can be combined with MSI and KM services to be made more "*cooperative.*"

Such platforms may provide a good basis for effective collaborative work and knowledge sharing on rapid design of customer-driven products. This new way of collaborative working allows EE to be fast and flexible enough to meet changing customer requirements, under reduced costs and time, being the key objective of the *New Product Design Model.* Generic and widely applicable, modular collaborative ICT platforms to support collaborative product/process design applicable for different design approaches are the main prerequisites for modern design in international setting.

For each product development phase the platforms include sets (library) of design tools appropriate for collaborative work in an EE. The selection of tools to be included has to be done based on specific EE needs.

For example, for the *product conceptual* phase, the platform has to include product design tools to support the main relevant areas that should be taken into account at the conceptualization of a new product such as:

- Functionalities
- Innovations that lead to new functionalities and improvements
- Expected life span of the product
- Integration of expertise and know-how inherited in the company from previous experiences
- General physical rules
- *etc.*

The library may include tools for

- *Innovation generation* (*e.g.*, TRIZ tools, see Chap. 1),
- *Voice of customer capturing and analysis*
- *Product life cycle analysis*
- *etc.*

The core services will support *on-line selection* of the most appropriate tools for this purpose (see the text to follow).

The critical problems of such an approach are tools/service interoperability and appropriate ontology relevant for different design phases.

The platforms have to provide various application SW services to support collaborative design in a global setting *collaborative application specific services*, *e.g.*, services to provide support in conceptual design of products in cooperation with customers, in collaborative design decisions, manufacturability investigations, *etc.* Table 5.4 provides a brief overview of some examples of such application specific services.

Platforms have to be open for easy generation of various other services to support different phases of collaborative design and for involvement of different actors (product designers and service providers, maintenance providers, shop-floor operators, end-customers).

The analysis presented in Sect. 5.1 has shown that the required *collaborative application specific services* have common "collaborative" actions which may be supported by a common approach. Therefore, the CWE solutions include so-called *Core Collaborative Services (CCS)*, addressing generic, application independent functionality to support collaboration, covering these common actions (such as resource discovery, collaboration traceability, knowledge provision, *etc.*). The generic CCS have to be combined with design tools and KM tools allowing for effective collaborative work and knowledge sharing among different actors.

Table 5.4 Collaborative application services for product/process design (examples)

Collaborative application specific services	Key objectives	SW functionality
Collaborative conceptual design	Support in definition/collection of customers/consumers requirements and other constraints/rules relevant for product design, turning into constraints and processing	SW support in identification of experts needed for defining requirements, building optimal team, re-use of previous knowledge on requirements, constraints, rules (identification of similar cases) based on traced collaborative work, *etc.*
Collaborative design decisions	Support in decisions regarding optimal design	SW support in identification of expert in worldwide EE, identification of previous "similar" problems/solutions, support in weighting of criteria for decision, *etc.*
Collaborative manufacturability investigations	Support in investigating customer/consumers requirements vs manufacturability of parts including, *e.g.*, tolerance analysis, *etc.*	SW support in identification of experts (*e.g.*, shop-floor experts with specific collaboration patterns due to time and space constraints), re-use of previous traced tolerance analysis sessions, *etc.*

Such an architecture to support design within an EE, partly developed within the InAmI project, is presented in Fig. 5.3 (InAmI 2006) and is a further elaboration of the architectures presented in Figs. 4.1 and 4.2. This architecture fits with the emerging *Collaborative Reference Architecture* (Expert Group 2006), presented in Sect. 4.2.4. The architecture includes all elements of the Collaborative Reference Architecture indicated in Fig. 4.2. It is open for different design tools and services.

It includes several layers:

1. *Information middleware.* This layer includes interfaces to different systems (legacy systems, distributed databases), to collect information on products/processes, needed for collaborative product design.
2. *Core Collaborative Services* (CCS). A generic set of services supporting collaboration among teams in a network which includes three subsets:
   - *Core Services for MSI.* These include services such as: resource discovery, team composition, collaboration traceability, selection of communication services – see Sect. 5.2.2.1.
   - *Core Services for knowledge sharing.* Product/process knowledge provision – see Sect. 5.2.2.2.
   - *Core Services to support design process.* Context–sensitive selection of design tools, results interpretation – see Sect. 5.2.2.3.

3. Service *orchestration layer*. This layer serves to combine different Core Services with different design tools. This layer provides the user functionalities and a synergic combination of CCS (allowing for different collaboration patterns). Therefore, orchestration layer provides the capabilities to build collaborative application specific services. The architecture includes orchestration instead choreography, since orchestration describes a process flow between services controlled by a single party which more corresponds to the cultures/approaches accepted by the industry (Peltz 2003).[3] The "uniforming" sublayer assures harmonization and management of CCS.
4. *Collaborative application specific services layer*. This layer includes a set of services to directly support teams in their design activities.

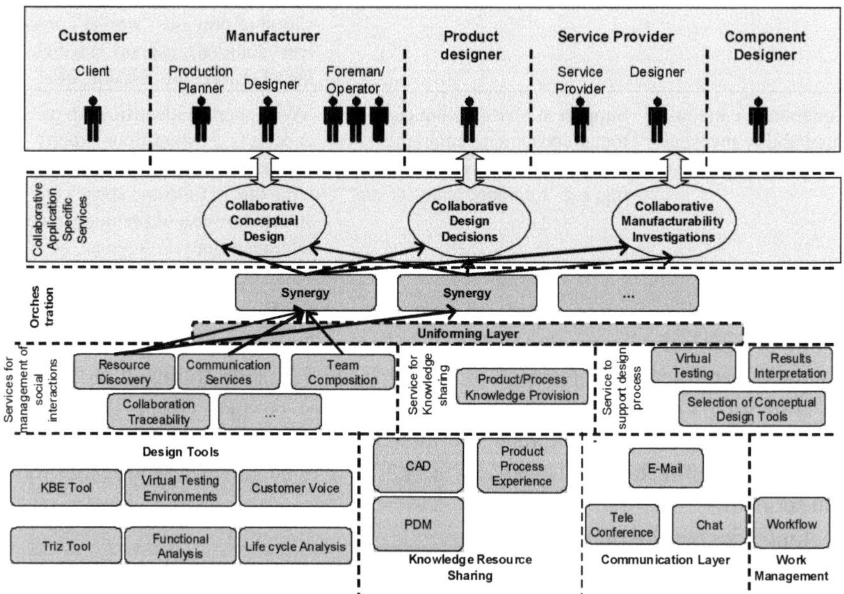

Fig. 5.3 Platform for collaborative product/process design

The architecture is implemented as an SOA allowing for full flexibility and effective instantiation of different collaborative design processes. The key issue is that design tools and the CCS support different patterns for collaboration among the teams: asynchronous, synchronous, multi-synchronous, local, distributed, *etc.*

[3] Choreography is concerned with interaction and conversation of web services, wherein languages, communication technologies, formal models along with techniques for operations like service compatibility determination or validity checking of conversation protocols is of interest. Orchestration is concerned with arrangement of several services to more complex functionality, wherein mainly service composition is of interest. Choreography and orchestration with web services are considered as the enabling technologies of web service-based process management – see http://events.deri.at/bpm2005/.

(as explained in Sect. 5.1.2). For example, CCS for team composition and resource discovery allows for an easy switching from one pattern to another, *e.g.*, from synchronous to asynchronous (Molli *et al.* 2002) which is specifically important for, *e.g.*, shop-floor teams. The platform specifically supports traceability and dynamic changes of collaborative design decisions within EE. An important aspect is that the services have to be accessible from any place at any time, allowing for full flexibility in collaboration. This means that platform supports mobile users.

5.2.2 Service Engineering Tools

Since the application services have to be dynamically updated/reconfigured due to frequent changes in collaboration needs and conditions and have to be integrated with other collaborative services and design tools, the platform allows for effective generation/update of such services. Service engineering tools for generation of *collaborative application specific services* either automatically update these services (*e.g.*, make changes in collaboration support based on tracing of the collaborative work) or allow users, non-IT experts, to generate/reconfigure application services by themselves. The tools support generation of application specific services which can be easily integrated in different environments (*i.e.*, with different information middleware, with different legacy systems) and which can be easily integrated with the existing design tools. This is fully in line with the *WEB 2.0* approach aiming to allow users to create their own services fitting their specific needs, as described in Sect. 4.2.2.

Following the generic use case for a generation of a new service, as described in Sect. 5.1.3, several groups of tools are included as indicated in Fig. 5.4. These tools serve to support generation of *collaborative application specific services* within the targeted architecture. In the text to follow, three examples of service engineering tools are briefly described (Stokic *et al.* 2007):

1. *Service composition.* The tool allows users to create a new application service, or edit/modify an existing application specific service. The functionality requires high-level access rights, as it implies access to sensitive information of an EE (*e.g.*, list of all users and respective rights). The tool allows users to define collaborative application specific service issues such as:

- The purpose of the service
- Text and structure for the help system to be used in the service
- Users who are allowed to use it
- Collaboration patterns which are supported
- Core services and additional functionality that are necessary to implement the application service
- The history of all actions related to creating and editing the service

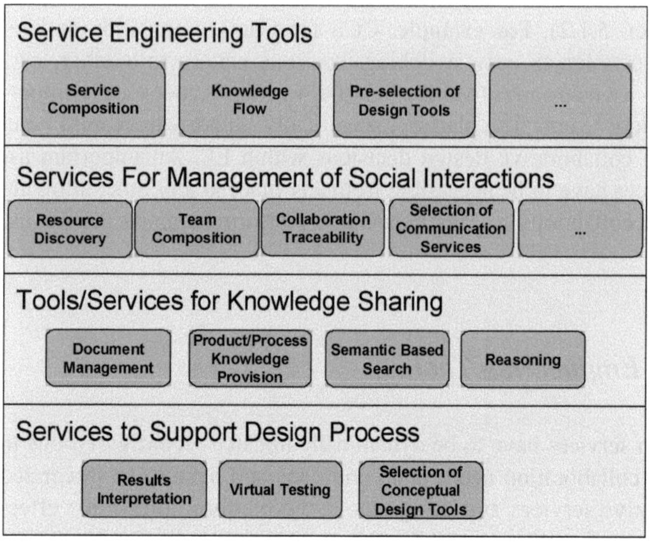

Fig. 5.4 Service engineering tools

2. *Identification of knowledge flow and definition of knowledge presentation and Graphical User Interface (GUI) for a collaborative application specific service.* The tool helps the user in defining which information is needed for the service, a selection of KM services/tools and auxiliary functionality and their potential use in the services. It provides a list of information/knowledge needed for the Application Service, together with sources from which this information can be collected. This tool supports the knowledge flow identification for the application services.

This tool can also support orchestration of services. As indicated above, the application service needs orchestration of Core Services and other tools. The Core Services are available as atomic web services which can then be composed/orchestrated in order to obtain composite collaborative application specific service.[4]

The tool has the functionality for presentation of existing knowledge objects in the system, filtered by the target user group's rights and the desired collaboration pattern, allowing a user to select which one is relevant for application service and for presentation of the available KM tool/functionality and other relevant tools (again, filtered by user rights and collaboration pattern) so that a user can select the ones needed for the application service.

This module also has the functionality to define the information which will be presented in the Graphical User Interfaces (GUI) for the application service; given a user or group of users and the available GUI, the designer may define which of

[4] An orchestration of Core Collaborative Services could be done by *e.g.*, Business Process Execution Language (BPEL) tools. The tool for identification of knowledge flow can help the BPEL process definition by providing BPEL partner links details (see Sect. 5.2.4).

them can be used by which user. The presentation of the available GUI can be made taking into account ontology of existing GUI. The ontology includes classes such as:

- GUI
- User group
- Information/knowledge object, and slots (properties of classes):
  - User rights
  - User device type
  - Collaboration pattern (some objects may not be visible with some collaboration patterns, or have restricted visibility)
  - Visibility, or viewing rights (an object is visible to a user group which has the proper rights for that; one of the GUI formatted for a desktop computer monitor is not visible to a user group which uses PDA).

The user's device type could be merged as a special case into user rights. The tool supports setting-up of an *ontology* appropriate for a specific EE. The problem of ontology can be addressed by applying approach for distributed set-up and maintenance of ontology (Kuczynski *et al.* 2005) as described in Sect. 4.3.3.

3. *Pre-selection of design tools.* The tool supports pre-selection of the design tools from the platform library for a specific EE during application service generation (taking into account specific EE needs and infrastructures, interoperability and IPR aspects), while the Core Services to support product design process provides on-line context-sensitive selection of tools (depending on the specific collaboration situation) – see Sect. 5.2.2.3.

5.2.2.1 Services for Managing Social Interactions

Following the generic use case presented in Sect. 5.1.3, the *Core Collaborative Services (CCS)* needed to support management of social interactions are identifed. The key layer includes four groups of Core Services for *Management of Social Interactions* (MSI) – generic set of services supporting collaboration in an EE. Table 5.5 provides an overview of the key Core Collaborative Services for MSI. This is not an exaustive list of Core Services, but provides a set of examples of relevant services.

These Core Services support different collaboration patterns. For example, the services for traceability of intra-enterprise collaboration strongly take into account results of work on traceability of project development and knowledge modeling (Bekhti and Matta 2002). These collaboration traceability services generate information/content on collaborative activities in the appropriate form for future re-use (context to be related to the applied ontology) and solve critical issues of handling different collaboration patterns, privacy and IPR issues (Tang and Molas-Gallart 2004); see Sect. 6.7. The critical problem is how to monitor and document knowledge-based activities without requiring people to explicitly document their activities, observing privacy and IPR issues.

Table 5.5 Core Collaborative Services for managing social interaction

CCS (groups)	Input/request	Output	Main functionality	Specific requirements
Resource discovery	Request for expertise	Appropriate and available expert(s)	Searching for expertise Check availability	Discovery of experts within EE available for solving the design problems, available for either asynchronous or synchronous collaboration
Team composition	Request for optimal team	Optimal team	Proposes team based on identified expertise	Enterprise rules, *etc.* Adjustable to different collaboration patterns and previous situations (collaborations)
Collaboration traceability	Request for tracing of the group work	Info. on the requested states of group work	Tracing of collaboration: – Continuous, or – Event driven, (event identification)	Specific requirements regarding security, companies' specific rules. Easily adjustable to different specific needs/constraints in an EE and supporting detection of the context of collaboration
Selection of communication services	Communication needs	Proposed communication means	Selects most appropriate means for the specific actor, pattern, *etc.*	Voice, text, drawings, images, videos, email, chats. Time zones aspects. Rules in EE Personal preferences

5.2.2.2 Services/Tools for Knowledge Sharing

Different tools for KM have to be combined within application service to support knowledge sharing among teams. The Core Services for product/process knowledge provision have a task to select the appropriate tool to provide knowledge (*e.g.,* for searching documents related to problems, *etc.*) taking into account different expertise of teams involved in collaboration (Stokic 2006) and addressing privacy and intellectual property rights (IPR) issues (Tang and Molas-Gallart 2004).

This is a set of services with access to product/processes (over information middleware), documents, stored user knowledge in knowledge repositories, legacy data bases, *etc.* These Core Services may provide, *e.g.,* "similar" problems to the one to be solved by an actor/group by applying case-based reasoning (CBR) and rule-based reasoning (RBR) tools (see Sect. 5.3.2.4).

Many tools for KM can be used for such services, but they often need to be upgraded by adding collaborative aspects. For example, for each actor within a collaborative product/process design the knowledge on his/her collaboration within different groups can be used in defining search criteria and/or weighting of different similarity criteria for CBR.

To this group of Core Services belong also services to support decision making which provide, *e.g.,* functionality to support weighting of criteria for decision and decision traceability, assuring hierarchy in decision making according to enterprise rules, *etc.* (Campos *et al.* 2007).

5.2.2.3 Services to Support Design Process

This set of services has to (on-line) support (context sensitive) selection of design tools and testing environments. The examples are services for selection of conceptual design tools and of *Virtual Testing Environment (VTE)* and Services to Support Results Interpretation (context-sensitive selection of tools for representation of product prototype).

The context has to be provided "on-line" by MSI Core Services: the Core Services for traceability of collaboration identify context (using ontology defined for a specific EE) of the collaborative situation. The selection can be done based on a "simple" context model.

Such a "simple" context model could be differently defined for each enterprise, including:

- Technical aspects such as:
  - Products/processes addressed
  - Technology addressed
  - *etc.*
- Collaborative aspects such as:
  - Temporal
  - Spatial aspects
  - Companies and actors involved
  - Expertise
  - *etc.*

Based on the identified context the services suggest the most appropriate design tools and VTE and representation of product prototype. A more detailed explanation on context sensitive approach in CWE, with more sophisticated context modeling and identification, is provided in Sect. 6.7.

5.2.3 Information Middleware

As indicated above, the collaborative architecture must have smooth interaction with underlying layers, in the first place with the information middleware, but also with the specific infrastructures in different EE environments. The platform has to allow easy integration with different legacy systems, communication layer, and existing knowledge infrastructure (*e.g.,* document management systems, CAD, drawings, *etc.*), to collect information on products and processes needed for collaborative services and product/process design. The platform is fully open for different information middleware.

5.2.4 Implementation Aspects

This section is dedicated to the solutions for implementing the platform for collaborative product/process design. The implementation view of collaborative platform is presented in Fig. 5.5 highlighting the layers of the platform. Conceptually, this architecture is following the design principles of the SOA approach (Correia *et al.* 2008a).

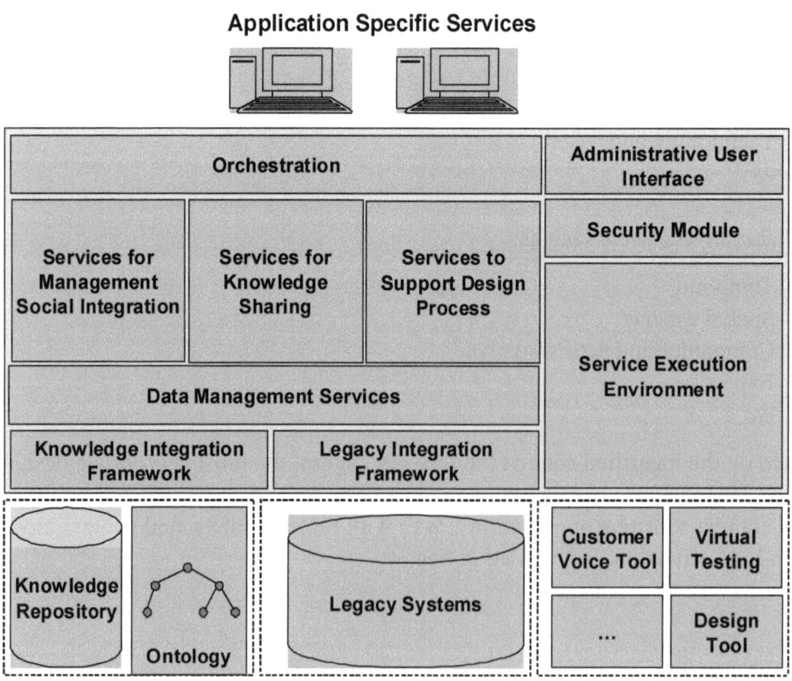

Fig. 5.5 Platform implementation view

Orchestration (top left). As indicated above, this layer provides an environment where collaborative application specific services can be developed with low effort. In this sense, the creation of applications can be performed by composition of fine grained services to coarse grained services or applications. Orchestration can be implemented:

- Manually (off-line)
- Over certain tools – automatically (on-line)

For the simpler implementation of the platform an off-line approach in which the combining of Core Services requires some additional activities by system operator/ICT provider/user can be applied supported by service engineering tools.[5]

Administrative user interface (top right) provides Web-based interfaces for platform administration. Using any Web browser, a user can set up user accounts, service execution environment, orchestration, *etc.* The interfaces remove the need to edit configuration files manually and let a user manage a system either from a console or remotely.

Security module (middle right) is a general module in charge of controlling the access to the platform as well as guaranteeing the integrity, confidentiality, and availability. An authentication and authorization framework with granular and manageable permission schemes is needed. From the authorization point of view, it addresses the management of users, groups and entities, as well as the definition of permission hierarchies, which can be grouped into roles. From the authentication point of view, the framework is able to authenticate users with existing credential repositories (*e.g.,* Windows®, UNIX® *via* PAM[6], LDAP[7]). To meet the requirements to solve ethical and privacy issues, the authentication and authorization framework also need to support audit logging.

Service execution environment (middle right): these services are a key enabler of SOA, a framework capable of managing all operational aspects related to SW services. This module can be seen as an execution environment which enables discovery, selection, mediation, and invocation of services. A SOA approach and more specifically, a Web service-based approach provides several advantages to the solution such as reusability, scalability and composability of services.

---

[5] Tools which allow simple workflow configuration as well as the definition of complex business processes using the BPEL may be applied (Trickovic 2006).

[6] PAM (Pluggable Authentication Modules) is an authentication method that gives administrators the ability to govern how users log on and authenticate themselves.

7 The Lightweight Directory Access Protocol (LDAP) is an application protocol for querying and modifying directory services running over the Transmission Control Protocol (TCP) and the Internet Protocol (IP).

Data management services (middle left) enable data transfer between platform and knowledge repositories and legacy systems and existing design tools and services. This execution environment manages all operational aspects related to services on the one side, and the integration frameworks on the other.

Knowledge integration framework (middle left) allows the integration of different knowledge sources into the system, such as knowledge repository and ontology. A user/developer can define the mapping between data retrieved from different knowledge repositories into a common format which can be understood by knowledge services on the server-side. The knowledge repository is the central source of information/knowledge for the collaborative product/process design system. Therefore, it must enable a proper structure to store all the knowledge related to application services. In addition, the common repository has to provide the means to model the products and processes within EE, serving as a basis for application services to be realized.[8] The repository and set-up tool allow each EE to model its products and processes as best fits their needs.

Core Services for management of social interactions (middle left): as explained above, these services are generic services (see Sect. 5.2.2.1), which may be set-up/customized for each specific EE using the service engineering tools (see Sect. 5.2.2).

Core Services for knowledge sharing (middle) are services (*e.g.,* for search, CBR, *etc.*) which can be used by actors to access "enhanced knowledge" (*e.g.,* services may constrain the search space to be used by existing searching functionality). They can vary depending on application scenarios but nevertheless are seen separately from legacy systems as they are an integrated part of the platform (see Sect. 5.2.2.2).

Core Services to support product/process design (middle right) are services to support directly (context–sensitive) selection of design tools and, *e.g.,* testing environments and support results interpretation (see Sect. 5.2.2.3).

Application specific design tools (bottom right): these are existing tools to support product/process design in different phases of the development process which can be combined with Core Services to provide application services and which may need to be updated to meet the requirements regarding interoperability. The tools to be used in the platform strongly depend on the application scenarios (see Sect. 5.2.1).

---

[8] There are several modeling methodologies (*e.g.,* ARIS, CIMOSA, GRAI/GIM, IEM, ENV 12204, *etc.*) which efficiently support business re-engineering and enterprise integration. In order to define the most adequate solution for an effective combination of knowledge management across the enterprise, which is flexible but also easy to understand, one of the most appropriate modeling languages to be used may be the European Standard EN ISO 19440. This Standard is called "Constructs for enterprise modeling", and was realized after a study of existing methodologies and languages.

Technical approach. The system functionality can be provided as a collection of services which can be accessed with an Internet browser (thin-client) or a common, locally installed, application (fat-client). The main advantage of this SOA approach is to provide the possibility to customize the system functionality (by adding/removing/customizing and orchestrating the services). As the integration with different legacy systems and availability on distinct platforms are key issues for such system, it is recommended to use open standards as much as possible. It is recommended (in order to allow high flexibility and support enterprises to take part in many networks of EE – multithreaded character of modern EE) that all SW tools for such a platform can be open source and the technologies be all open standards, *e.g., eXtensible Markup Language (XML), Hypertext Transfer Protocol (HTTP), Simple Object Access Protocol (SOAP), Universal Description, Discovery and Integration (UDDI), etc.*

5.2.5 Application Scenarios

In the text to follow, two typical application scenarios of the presented platform are briefly described. The described above platform has been applied in several industrial companies partly in the scope of several research projects (InAmI 2006; Correia *et al.* 2008b). The presented applications address real industrial environments in companies in UK, Turkey, Germany, and Poland.

Scenario 1: Collaborative vendor-customer conceptual design. The scenario addresses collaborative product design between Company Y – customer and Company X – equipment provider. Both companies are SME. The equipment vendor X from UK is developing and marketing complex automation and robotics (A&R) equipment to industry. The company X is a part of a world wide network of suppliers of engineering services with extremely high diversification of solutions and applications offered (for different sectors). X applies a new business model with their partners, in which they not only engineer new A&R solutions and sell them to customers, but also take over full responsibility for A&R operation, providing new types of services. To apply such models, company X uses ICT solutions which allow for much tighter collaboration with all geographically distributed suppliers and customers, especially in initial design phases. Therefore, X uses the ICT system described above to support virtual design teams to develop new A&R solutions for customers all over the world. The main issue is to involve both employees and customers in the product/service design process. Thanks to its well-developed after-sales service, which includes remote control and maintenance, the company X has the possibility of implementing customer feedback for innovation.

The platform is successfully used with one of the new customers, *i.e.,* the company Y from Turkey, with whom stronger business relations are established. The customer Y as a relatively young manufacturing company (producing temperature control elements) intends in the near future to improve its manufacturing process with a new automation line with a number of new (specific) automation systems and machines, to be specially designed by the vendor company X.

Scenario 1, therefore, focuses on collaborative gathering of customer requirements and conceptual design of automation solutions between the UK company X and the customer Y in Turkey. The application shows how this new collaborative design environment can be used to build new business models with customers and vendors. Since both vendor X and customer Y are involved in several EE networks (multithreaded aspects), the tools to reconfigure easily collaborative application specific services are of special importance.

The benefits obtained by usage of platform are manifold: the time and efforts needed to develop new A&R solutions are drastically reduced by at least 25%. The new business model, facilitated by the platform, brings obvious benefits regarding the trust and long term relations to customers.

Scenario 2: "Consumer-driven" manufacturability at suppliers. This scenario addresses the key problem encountered in the automotive industry supply chain (distributed over the globe), i.e., the collaboration between original equipment manufacturer (OEM) and first and second tier suppliers during the design phase of new product variants and related components. In the next years the company X (OEM) from Germany needs to develop many new models with numerous variants. Therefore, they establish new relations with their suppliers. One of the partners in a supplier group, the company Y (from Poland), offers full design and manufacturing of components to the OEM.

The collaborative application specific service for manufacturability investigations is used, addressing collaboration in releasing new product variants according to the consumer requirements, among:

- Product design team at company X (OEM)
- Design teams at both company X and company Y
- Manufacturing (shop-floor) teams at component supplier company Y

The process chain is the development and manufacturing of car seats. In this case the company X (OEM) uses functionality of collecting consumer requirements, conceptual design, and design control while the supplier group covers the detailed design and manufacturing. In order to fit not only engineering constraints but also cost constraints, the seats are designed on the basis of car class dependent modules which must be varied due to the specific car model, manufacturing constraints specific for plant, automation degree in the specific plant, car delivery country dependent legal regulations (legal constraints, *e.g.,* USA), and specific consumer requirements. The critical issue is how to achieve collaboratively compromise between consumer requirements (to be efficiently transmitted from OEM to suppliers) and manufacturability. Therefore, a service for collaborative design decisions is also applied. The benefits for both company X (OEM) and company

Y (supplier) in terms of reduction of effort and time for design and investigation of manufacturability aspects are considerable, leading to more efficient manufacturing of newly designed car seats (less problems in manufacturing of parts).

5.3 ICT for Collaborative Innovation Management

As indicated in previous chapters, the modern approaches for product/process innovation in industry require involvement of many different teams in EE context. In the previous section, collaborative work on product/process development and ICT support for such collaborative work in an EE have been examined. In this section, management of collaborative work on innovation in product/process development is addressed. The key problem of innovation is how to involve effectively all stakeholders, both organized and non-organized teams (*ad hoc* groups and wider communities, *e.g.*, RTD, customers), in such an overall innovation process, and organize their efficient collaborative work. Specifically the intention of this section is to examine how ICT may support such collaborative work on innovation.

The objective of this section is to explore how Collaborative Working Environments (CWE) for collaborative product/process design, presented in the previous section, can be extended to support collaboration in industry within an overall innovation processes, *i.e.*, to provide a *computer aided innovation (CAI)* system to support collaborative work on product/process innovation. The platform is primarily intended to support collaborative work on innovation within organized teams in industry in EE context, but it is extendible to support collaboration between teams in industry and *ad hoc* groups and wider communities.

5.3.1 Innovation Process Baseline

As indicated in previous chapters, innovation processes can be organized within an EE in various ways. The innovation in industry can be fostered in one of two ways

- By requiring innovative solutions of identified products/processes problems and improvement potentials
- By directly and continuously collecting ideas from all involved actors in an extended enterprise (independently of the identified problems)

As explained in Chaps. 2 and 3, the innovation approaches observed a span from classical incremental innovation approach up to the systemic innovation in industry. For all modern innovation approaches, collaborative application services are needed to support, *e.g.*, the collection of ideas and a collaborative work (both synchronous and asynchronous) around the gathered ideas. The analysis has shown that the collaborative innovation processes have common collaborative activities (*e.g.*, team building) which may be supported by common approaches, as described in Sect. 5.1. These common actions can be supported by so-called Core Collaborative Services – as explained in Sect. 5.2.

In the text to follow, an innovation process approach is described in detail as a baseline for a definition of ICT support to innovation processes in industry, *i.e.*, for a definition of core collaborative service (Sorli *et al.* 2006a). However, the ICT solutions described are applicable for innovation approaches that may differ from this innovation process.

The proposed baseline is thought as a holistic process of innovation (Zlotin and Zusman 1999), which means that an "idea" will undergo a complete cycle in order to be collected, documented, classified, and used in the ICT system. Ultimately, *ideas turn into innovations*, which is one of the main objectives of the system. This section provides a rough model of the life cycle of an idea. Figure 5.6 shows the path that an idea undergoes in an EE.

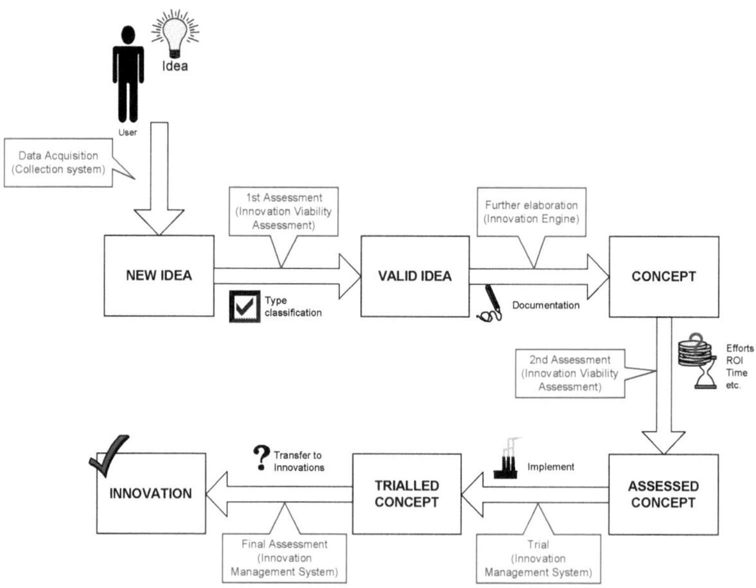

Fig. 5.6 Innovation life cycle baseline (AIM 2005; Sorli *et al.* 2006a)

The rationale followed to support innovation process in EE is based on the following assumptions:

- The ideas for product/process innovations have to be collected throughout an EE in order to use all potentials for innovation available in an EE

- The ideas are provided either as possible solutions to identified problems in products/processes or as potential improvements of products/processes
- The ideas generation and gathering can be stimulated by provision of knowledge on:
  - Problems related to processes/products
  - "Similar" ideas on products/processes
  - All other available knowledge on products/processes
- Ideas have to be effectively assessed to select those which are most likely to lead to innovative solutions (process or product innovations)
- Ideas may need to be combined to achieve innovations

This life cycle is the basis of the innovation process, containing the activities to be realized to achieve innovations in an EE. The life cycle starts with data acquisition, where ideas are collected using an appropriate graphical user interface, accompanied by knowledge acquisition methods.

Next, a first assessment of the *new ideas,* with the purpose of making a rough classification is taking place. This classification will be an identification of the idea type, according to the information that it contains: improvement, potential cause, action or new product/process. The main objective of this first classification is to attribute a type to each *new idea*, enabling its fast identification by the appropriate staff members of the company.

With all the ideas classified by type, a responsible staff member will develop *valid ideas* further, by first collecting any additional information that might be relevant for the *valid idea* and then elaborating it in more detail.

All the information can be useful to enable the best possible assessment. This step also includes relating the idea to other ideas, innovations, and information stored, such as products, processes, problems, causes, actions. The result of this step is a more elaborated idea: *concept.*

The company's staff members responsible for ideas' evaluation will realize a detailed assessment of each *concept,* with the objective of supporting a decision of trying or not the idea, *i.e.,* implementing it. Several issues must be considered here, such as material, machines, staff members, implementation cost, profit, efforts, and return on investment (ROI). The result of the assessment will be documented in the repository, together with the *concept*, defining an *assessed concept.*

If the result of the assessment expresses an expensive and unworthy implementation, the *assessed concept* will probably not be implemented, and this has to be documented. It is then possible to keep the concept in the repository to reuse part of its information, or delete it. When the assessment provides positive results, the *positively assessed concept* is tried, and the complete development process is documented in the repository. The most important part of this documentation is the result obtained from the trial implementation, which expresses the success of the concept or not, and defines a *trailed concept.*

The documentation of the *trailed concept*, collected until this step, enables the final classification of the initial ideas. Based on the assessments and the trial implementation it is possible to identify if the idea is successful and therefore constitutes an *innovation.*

In this line, the ICT platform for *computer aided innovation* (CAI) has to support the collection of innovative ideas and relevant knowledge throughout the EE for new and existing process and product developments. These ideas and knowledge will later be developed in a collaborative way fostering industrial innovations, as team work will be enhanced by co-operation between manufacturers, customers and suppliers by means of the Internet facilities provided by the CAI platform, "accelerating" innovation into the market.

The CAI platform to be described focuses upon the initial phase of innovation process: generation and collection of ideas, through their assessment and structuring up to their delivery to the product/process design teams which will turn these ideas into innovation. The platform and services for innovation management have been developed and applied within the research project AIM (2005).

5.3.2 ICT Platform to Support Collaborative Innovation Process

The *computer aided innovation (CAI) platform* described in this section is a generic and widely applicable modular collaborative ICT platform to support work on innovation, analogue to the one described in Sect. 5.2.1 for collaborative product/process design. Actually, this CAI platform is an extension of the platform presented in Fig. 5.2. The platform has to provide various collaborative application specific services to support innovation in industry, especially those enabling the teams in shop-floor to be involved in innovation processes, *e.g.,* for process improvements of manufacturing processes or improvements of product manufacturability, *etc.*

The platform is open for various services to support innovation as well as to the involvement of different actors. The examples of application services following the above presented innovation process baseline are:

- Collaborative problem solving, either product problems or process problems
- Collaborative process improvements: collecting and assessing ideas to improve processes, developing new concepts, *etc.*
- Collaborative product improvements/design *etc.*

As explained above, application services use the *information middleware* which provides information on products/process/production units needed for collaborative work on innovation, *i.e.,* the information middleware includes interfaces to products/processes and other systems.

The proposed platform is presented in Fig. 5.7. It is again an instantiation of the general platform presented in Fig. 4.1. The platform includes three layers:

- Core Collaborative Service (CCS) layer
- Orchestration layer
- Collaborative application specific services layer

The platform also includes innovation repository and a set of tools to build services to support collaborative innovation.

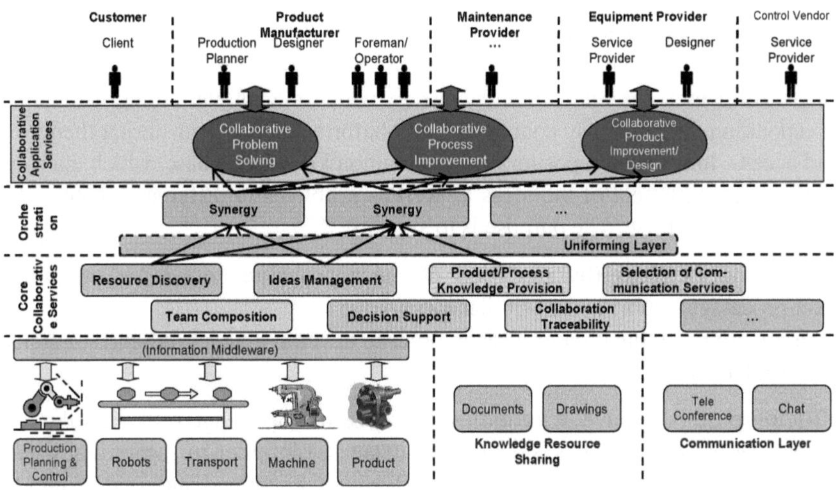

Fig. 5.7 ICT platform to support collaborative innovation

Some typical users of the platform are:

- *Designers, both product and equipment designers.* To utilize feedback to improve existing products and also to work on new design ideas of their own. They will be expected to manage the design process. Having better access to more knowledge should reduce time to market and result in better products and an improved product range.
- *Shop-floor and technical support personnel from the manufacturer.* To provide feedback on product designs, report problems in the manufacturing of the products and provide suggestions on new products or product improvements.
- *Management.* To provide their ideas, to monitor the innovation process, to support decisions making on ideas/innovations to be elaborated/tested, *etc.*
- *Customers.* To provide suggestions and knowledge and to develop their ideas about improving current products or new products they would like to have.
- *Field operatives from the equipment provider* (*e.g.,* maintenance, service provider). To provide feedback from ideas, suggestions, knowledge, and to de-

velop further their ideas about new products they have discussed with the customer.

- *Suppliers* (*e.g.,* component vendor). To develop knowledge and to provide feedback on the designs that they have been asked to provide their expertise on (*e.g.,* where the new products will require material, parts and components from them or new equipment to manufacture the new products), as well as to provide novel ideas.

5.3.2.1 Innovation Repository

The platform includes *innovation repository* which allows classifying ideas and corresponding data in a way common to all platform services and storing them for rapid access. Innovation repository is the common knowledge base, which enables an effective attachment of the ideas as well as problems to different "objects" in an extended enterprise. The purpose is to support:

- Collection of ideas (by provision of the appropriate knowledge on product/processes, problems and similar ideas)
- An effective combination of ideas in order to develop innovations and solutions to problems

The purpose is to have a "centralized" source of knowledge (although the repository can be organized as a distributed database) in order to assure that ideas coming from different sites of EE can be effectively combined and used (Campos *et al.* 2004).

Repository structure. This repository classifies innovations using an "innovation" meta classification, and stores them for rapid access. This includes:

- *Product/process knowledge base.* This knowledge base includes all relevant information and models as well as experience-based knowledge of products and processes. This knowledge base includes data on:

  – Products
  – Processes (including manufacturing processes, logistics, but also design, marketing and other business processes)
  – Production units (*e.g.,* machines, transport lines, robots, tools, *etc.*)

  Additionally knowledge base includes data on

  – Involved technologies,
  – Human resources
  – *etc.*

  Such a knowledge system must be related to a number of other legacy systems in a company. In some companies, such a knowledge system may already exist and therefore the *computer aided innovation* (CAI) system has to be connected to it (raising problems of interfaces and extraction of knowledge needed for CAI services/tools). However, in a number of industrial companies, such a knowledge system does not exist in an appropriate form. Therefore, the products/process knowledge system is a part of CAI but in the case that such system already exists, CAI must re-use it.

- *Problems/potential improvements repository.* This repository includes knowledge on problems and potential improvements regarding products/processes. This covers knowledge on problems identified, their reasons, and/or actions that were used to solve them in the past. These also include so-called *actual state items* which represent current (real-time) states of the products/process/production units related to the identified problems. For CAI system the problems which have not been (appropriately) solved and which ask for innovative solutions are important. Similarly, as for a product/processes knowledge base, in some companies such a repository with problems may already exist (or be a part of a product/process knowledge base), but a number of them will not include all relevant knowledge.
- *(Innovative) ideas and innovations.* All ideas and innovations are stored using the meta classification. The overall meta classification of the (innovative) ideas and innovations is defined as a basis for all CAI components.

The innovation repository has to allow easy modeling of products/process in different companies. Special attention has to be given to a definition of an appropriate classification for different specific products and processes within a specific EE context. The repository model is presented in Fig. 5.8.

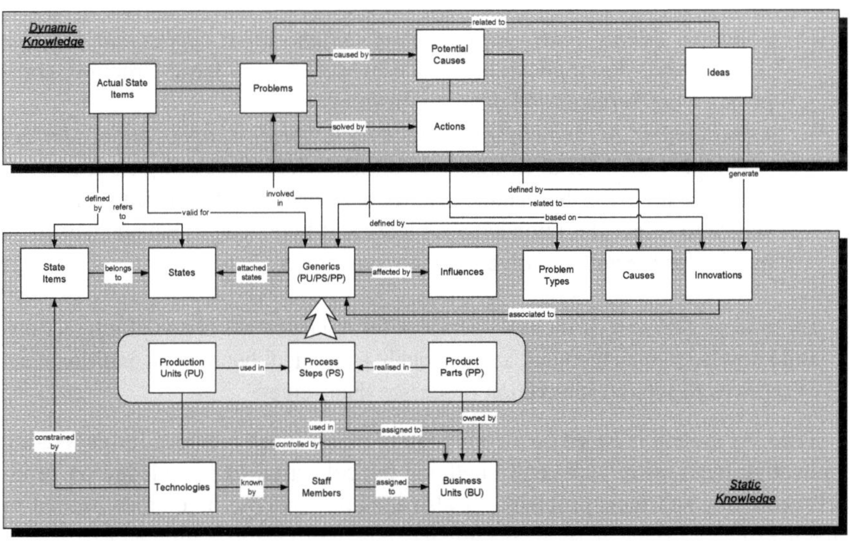

Fig. 5.8 Innovation repository maintenance model

Repository maintenance. Those services for the set-up and maintenance of the repository are quite important. Set-up enables each company to model the repository as it best suits their specific needs. This means that each company may within set-up define the structure of their products and the structure of the product features, *etc.* The classification of ideas is also adjustable to specific user needs, *i.e.,* the set-up allows a definition of a classification appropriate for the user. The set-up also allows introducing static, standard information in the repository.

5.3.2.2 Tools to Build Services for Collaborative Innovations

Since the services have to be dynamically updated and have to be integrated with other collaborative services, there is a clear need to provide the described platform allowing for an effective generation/update of such services. Therefore, means/tools to generate efficiently different application services for collaborative work on innovation among groups in an EE are needed (Expert Group 2006).

The described *computer aided innovation* (CAI) system follows SOA architecture, enabling an easy extensibility, robustness and customization, and supporting the activities identified in the idea life cycle.

The main tools are intended to support development of collaborative application services which can be easily integrated in different environments and with the existing applications (*e.g.,* existing systems for innovation, concurrent engineering tools, CAD/CAM systems, *etc.*). Similar to the platform for product/process design, as explained in Sect. 5.2.1, the tools include:

- *Tool set.* It supports:
  - *Creation, editing and composition of collaborative application services* which create the application service by defining its basic structure, composing Core Services and other applications
  - *Identification of knowledge flow in collaborative application services* which provide a list of available information/knowledge objects and set of available tools for the management of knowledge and allows a service designer to select the knowledge objects and knowledge management tools needed for the collaborative work on innovation

- *Set of Core Services.* It includes services for management of social interactions and knowledge management, as previously explained, but also an additional set of services to support innovation processes themselves.

The CAI system is based on a combination of advanced methods for generating innovative ideas with "classical" methods for collection of knowledge on products/processes and their problems. It also includes specific ontologies needed to enable efficient exchange of ideas/knowledge between actors within an EE.

5.3.2.3 Core Collaborative Service for Innovation Management

The platform comprehends several Core Collaborative Services (CCS) to support an innovation process directly, following the proposed innovation process baseline – see overview in Table 5.6.

Each phase of the proposed innovation process baseline can be supported by a set of services. In the text to follow these services are briefly described following the numbering from the table.

Table 5.6 Core Services for innovation management

No	CCS	Input/request	Output	Main functionality	Specific requirements
1	Collection of ideas	Problem defined, request for ideas	Ideas collected	Presents problems, collects ideas from different actors	Different representation of problems and ideas for different expertise
2	Collection of knowledge		Problem described	Collects information on problems – products/processes	Different ways to describe problems for different expertise of actors
3	Innovation Engine:				
	a) Innovation generator	Ideas defined (valid idea)	Structuring of ideas, interconnecting, concept		
	b) Problem solver	Problem defined	Proposes possible solutions		
4	Innovation viability assessment			Performs first and second assessment of ideas	
	a) First assessment	Idea defined	Valid idea		
	b) Second assessment	Concept defined	Assessed concept		
5	Innovation management	Accepted concept	Innovation	Planning and monitoring the use of knowledge	

1, 2. Collection of innovative ideas and product/process knowledge. This set of services provides the means to collect ideas efficiently, but also to collect knowledge on product and process problems for which ideas are needed.

The subset of services for handling collection of ideas covers the following functionality:

- Insert new ideas in the common knowledge base (CKB)
- Modify ideas already stored in the CKB

- Delete ideas
- Search ideas using basic criteria (responsible user, date, generic involved, problem related)
- Obtain ideas similar to a selected one, using case-based reasoning (CBR)

In order to support users optimally in introducing their ideas, different Graphical User Interfaces (GUI) to add or edit ideas are available. Any user can choose the most appropriate one and this choice can be modified at any moment (Campos *et. al.* 2004).

The subset of services for collection of problems/requirements covers the following functionality:

- Insert new problems/requirements in the CKB
- Modify problems/requirements already stored in the CKB
- Delete problems/requirements
- Search problems/requirements, using basic criteria (responsible user, date, generic involved)

The specific GUI to add/edit a problem is displayed in Fig. 5.9.

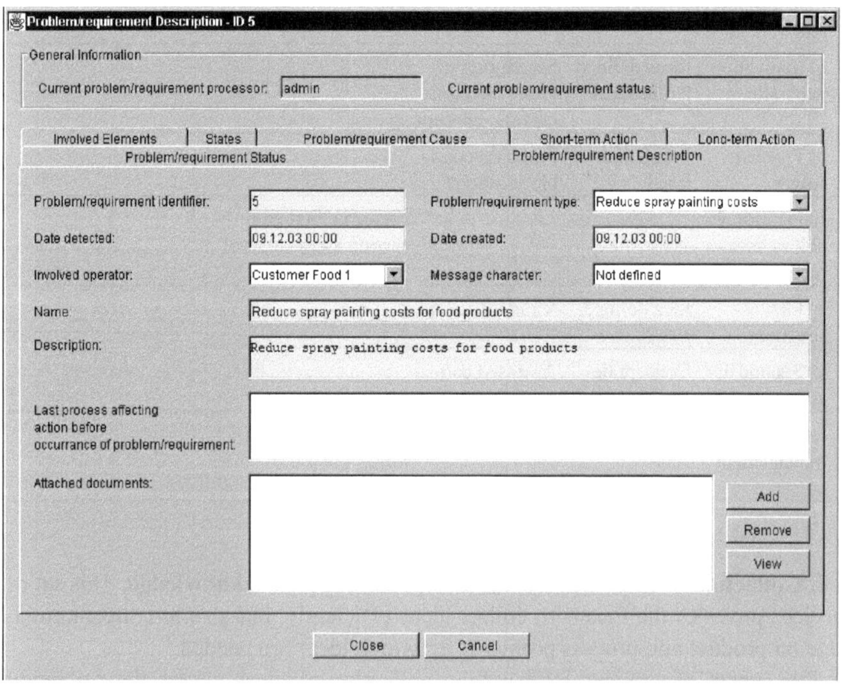

Fig. 5.9 Problem description

The problems/ideas collected *via* this GUI are stored in the knowledge base and related to products/process models in the knowledge base (see Sect. 5.3.2.1), which allows for their efficient re-use. The ideas are further processed and assessed using other modules (as will be explained in detail in the text to follow) and stored in such a way that they can be easily reused for motivation of people to innovate or to be combined later with new ideas.

The key issue of this set of services is provision of "similar" ideas and/or problems serving as means to support generation of new ideas. As explained in Sect. 5.3.1, the main assumption is that by the provision of the knowledge on "similar" problems and ideas, and other relevant knowledge on processes/products, the people in an EE will be inspired to generate the new ideas either as solution of problems or as potential improvements. The key problems of provisions of "similar" ideas and problems are discussed in Sect. 5.3.2.4.

3. Innovation engine. This set of services provides a systematic methodology support for the development of ideas into innovation concepts, by sharing and working on these ideas in a structured framework. The ideas collected by the services described above and stored in the repository can be further developed by the users. *Theory for Inventive Problem Solving* (TRIZ) methodology serves as a baseline approach (Kohnhauser 1999) for these services, where the in-depth analysis of technical requirements and (manufacturing) failure situations is performed, structured knowledge is delivered, and graphical aids for team working and creation of concepts from validated ideas (see Sect. 5.3.1) are provided.

Developing engineers and failure analysts may reason by analogy when facing new situations or problems. Many non-expert failure analyses reach the wrong conclusions simply because they fail to define the system adequately or consider a case in isolation from a history containing other related or similar cases. Therefore, within these ICT services a combination of methods is applied in order to make use of past knowledge and practices to overcome problems, even in different areas of application. The necessary knowledge to implement the reasoning methods exists in the innovation repository, either as innovations, ideas, or product and process knowledge (included in the product and process models).

The innovation engine services consist of two subsets:

- 3a. Subset to support the development of innovative ideas: innovation generator
- 3b. Subset to search for solutions to problems: problem solver

3a. *Innovation generator.* This is the innovation engine for the product/process improvements work-flow. This facility provides a structured means for supporting the development of ideas into new or existing product/process improvements. The ideas/innovations collected and stored in the repository will be further used for "innovative developments," understood as any matter that must fulfill specific requirements and requires an inventive solution. This will be the means by which raw, creative ideas can be organized and developed by sharing and working on these ideas/innovations in a structured framework.

3b. *Problem solver.* This is the innovation engine for the problem solving work-flow. This is the facility that provides a possible solution to a problem detected, *e.g.*, within the manufacturing process, based on previous similar problems or situations stored in the repository, the products/processes knowledge and ideas/innovations stored in the repository. It also supports the decision on actions for solving the problem.

The main difference between the subsets is the TRIZ-based methodology scope: while that for the innovation generator is focused on developing ideas into innovative concepts, the problem solver one is focused on depth-analysis of problems in (manufacturing) facilities and the generation of concepts of solution for these problems. Both subsets share the same knowledge and information nurturing means, namely ideas for the innovation generator and problems for the problem solver.

4. Innovation viability assessment. These services provide a structure (based on rapid consulting within the company of evaluation of developments and risks, combined with a multi-criteria decision support) to assist users in assessing the feasibility of new ideas at the collection stage, and innovation assessment facilities for design teams. It is important to focus on feasible, good innovative knowledge and develop this.

The innovation viability assessment is the set of service oriented to assist users in assessing the feasibility of new innovative ideas, developing them into concepts. Innovations that cannot be turned into reality for technical, commercial or socio-economic benefits are of little use. The assessment is realized in two distinct steps, both supported by the system (see Fig. 5.10).

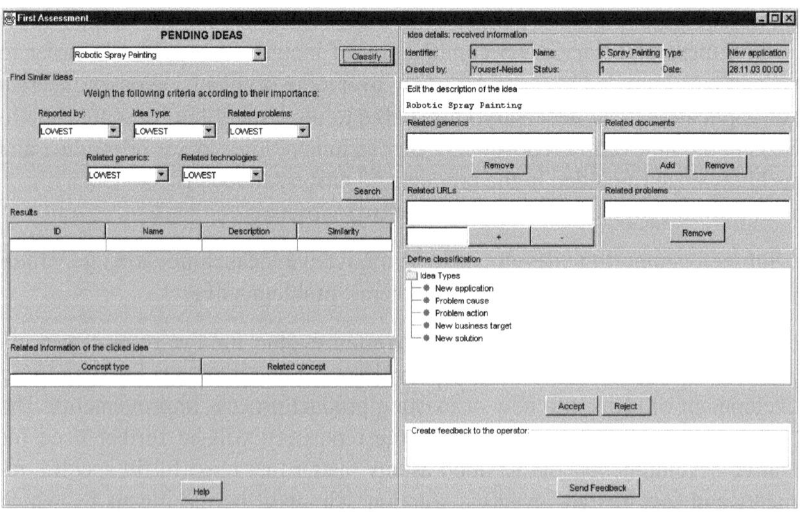

Fig. 5.10 Assessment screen example

4a. *The first assessment* consists of a rough classification, identifying the type of the "ideas" collected throughout the extended enterprise, classifying them according to a scheme defined during the set-up phase of the system (*e.g.,* improvement to product/process, cause for problems identified, new product/process, *etc.*).

4b. *The second assessment* comprehends a detailed study of the new developed "concept" before implementation. This assessment is based on strategic policy, including:

- Technical viability
- Implementation cost
- Materials to be used
- Equipment
- Profit expected
- Corporate efforts
- Return on investment
- *etc.*

These services are nurtured by the innovation repository where all proposed ideas are stored. The ideas collected must be related to the problem or product/process that they intend to improve, or to other ideas to which they are related. Later, these ideas are filtered, focusing on commercial or socio-economic interest, possibilities for turning them into reality, *etc.*

5. Innovation management system. This is a means of providing an efficient way for planning and monitoring the use of the innovation knowledge during design activities and a structured delivery of the innovations/ideas to the process and product design teams. For that purpose, these services use all other services. For example, innovation management system uses collection services when collecting a new idea to start the life cycle. However, these services are also able to deliver information to the final users, in the way of feedback to those who participated in the life cycle and in the way of statistics. For example, innovation management system provides statistics on system use and success (*e.g.,* new ideas, status of innovation process, users, number of innovations, number of problems solved). Innovation management system can also be seen as an orchestration layer in the platform, but could be implemented as a separate set of services.[9] With such workflow solutions, processes can first be modeled, using process modeling design tools and then implemented. A process model is designed to keep track of the life cycle of an idea.

---

[9] For example, innovation management system may use workflow solutions, such as FORO – a commercial workflow engine from ATOS Origin (AIM 2005).

5.3.2.4 Reasoning Approaches

The platform combines existing tools and technologies in a new way. The platform fosters creativity by combining classical reasoning methods, *case-based reasoning (CBR)* and *rule-based reasoning (RBR)*, which focus on the company's business objectives with an innovation supporting method, *Theory for Inventive Problem Solving* (TRIZ), and graphical aids for combination of concepts, within the context of specific products/processes, formalized by the use of continuously adapted ontologies.

As explained in Chap. 1, TRIZ methodology refers to the use of past knowledge to overcome problems and both RBR and CBR use past information, gathered in rules or cases, to reach a result. The necessary knowledge to realize these reasoning methods is provided in the system (in the knowledge repository), either as innovations, ideas, or product and process knowledge. The reasoning methods are very adequate to present a possible solution for problems (*i.e.,* new ideas or previous solutions), as the system contains information on past experiences (Thie and Stokic 2001). All three reasoning approaches are used to combine the ideas into innovation concepts by selecting those that may fit together based on previous appropriate combinations of these ideas.

CBR, for example, is used in three different services of the presented *computer aided innovation* (CAI) system:

- Collection
- Innovation viability assessment
- Innovation engine

For each of these services, cases have to be built, with the information that will be considered to evaluate similarity. The functionality can be used:

- For problems: to identify the respective causes
- In ideas: to support an effective collection of knowledge and appropriate assessment

The system may not need only to learn possible problem/idea situations and their causes but also to learn possible descriptions of problems or ideas. For this reason, the common structure of cases has to be more generic than in conventional CBR systems.

Similarity of ideas is based on five fields of information:

- Idea type
- User who reported the idea
- Involved processes, products and production units/tools
- Involved problems
- Involved technologies

Similarity of problems is based on the following information:

- Processes
- Products
- Tools involved in the problem
- Actual state items defined for the problem (values)

When two ideas/problems are analyzed to ascertain their similarity, each of the fields is compared to check if the respective information is the same in both cases. The comparison of each field provides a result (in percentage), which is afterwards computed in a total percentage, representing the similarity of the two ideas/problems. The priority given to each field is a value within *Ignore, Lowest, Below Normal, Normal, Above Normal* and *Highest* (see an example of interface for this search in Fig. 5.11). Furthermore, it is possible to filter ideas on the similarity comparison by selecting specific status that should be ignored (*e.g.,* the user can choose to ignore in the analysis of the ideas marked as *Invalid*).

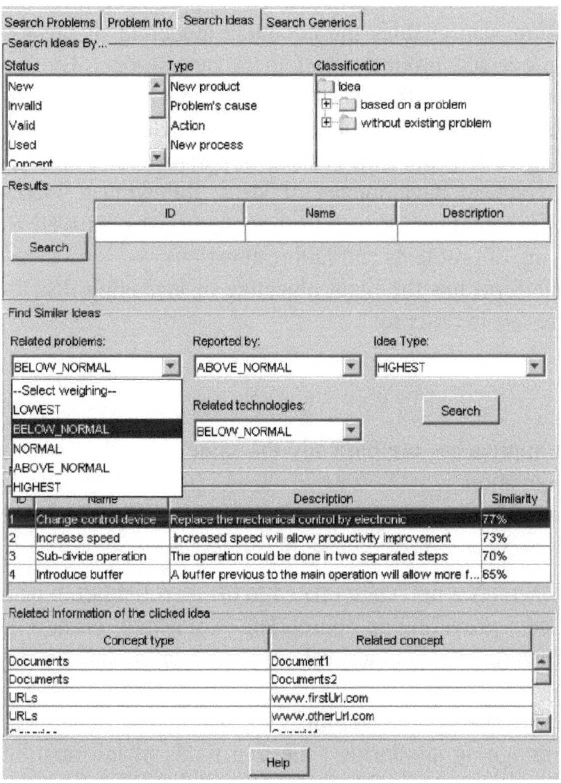

Fig. 5.11 Searching ideas screen example

As explained in Sect. 4.3, one of the key aspects of such a system is *ontology* which enables sharing and reuse of knowledge and reasoning behavior across different users, domains, and tasks. Ontology can be seen as complementary reusable components to construct knowledge bases. *Computer aided innovation* (CAI) is used by a wide spectrum of users with different technical backgrounds and sometimes with different languages. Therefore, it is highly important to have a common "dictionary" to avoid misunderstandings. By adding and retrieving innovative ideas, it is essential to ensure that certain words or keywords really mean what the users want to say.

5.3.2.5 Overall Platform Operation and Implementation

The main features of the described *computer aided innovation* (CAI) platform are obvious from the above description of the key services. The platform enables users along the EE to introduce ideas and report problems and provides functionality to validate these ideas. It also supports the assessment of the ideas developed in terms of technical viability, resources, costs, benefits. To support effectively the innovation process the platform also enables the complete modeling of the EE (*i.e.,* the departments, staff, processes, products, customers, innovations). By this the platform supports appropriate and efficient structuring and classification of ideas and problems. As explained in Sect. 5.3.2.4, the platform includes extensive search services for ideas, using various attributes as search parameters. It specifically supports users in the technical development of ideas, following a TRIZ-based methodology, for in-depth analysis of technical contradictions, as well as for in depth-analysis and solving of problems and failure situations.

In summary the overall platform has the main objective of managing distributed knowledge enabling its re-use in order to:

1. Innovate
2. Solve problems

It could be said that both approaches are basically the same since the *solving problems* is to be done *via* innovation.

The basic way of operation, following the proposed innovation life cycle baseline (Fig. 5.6), is:

1. Anyone in the company generates a new *idea* related to his own field of expertise and input it to the system providing an appropriate explanation on how it could be used, what is it intended for, and any relevant information that he/she will find interesting.
2. The *idea* is validated through the system administrator. He/she will complete all required information (assigning similarities, the five fields of information mentioned above, *etc.*); if needed he/she will ask for more details to the person that has generated the idea and finally will store it in the system.
3. Someone searching for an *idea* to either innovate or solve a problem, will realize a comprehensive search throughout the system with the help of the similarities (percentage) rates.

4. Selected *ideas* will be treated by the *innovation engine* in order to transform *ideas* to *concepts.*
5. *Concepts* are then assessed through the *innovation viability assessment* services.
6. Filtered *concepts* can then be trailed in real operative environments.
7. Successfully *concepts* finally become *innovation.*

For example, a specific problem in the production machine (see Scenario 2 in the text to follow) can be analyzed through an idea searching for similar problems in similar installations in other manufacturing plants of the same company, in other equivalent machinery in other industries of the sector, or many other possibilities depending on the amount of information available in the CAI platform (step 3).

Most suitable ideas will be treated by the innovation engine (step 4) and be converted into concepts; assessed through the innovation engine (step 5) and finally implemented into operation (steps 6 and 7).

Platform implementation. SOA approach is used, allowing for full flexibility and effective instantiation for different innovation processes. The platform is built as an Internet-browser based GUI application, a Wiki-based application for Core Collaborative Services and a centralized application server to achieve more sophisticated functions based on Enterprise JavaBeans™, similar to the one presented in Sect. 5.2 (Stokic 2007).

5.3.3 Application Scenarios

The platform and tools presented have been applied in different ways to support collaborative work on innovation, depending on specific EE needs (AIM 2005, InAmI 2006). In the text to follow, two application scenarios in real industrial environments are briefly presented (AIM 2005).

Scenario 1: Innovation in service and engineering in a medium-sized company. Scenario 1 focuses upon innovation in product development and product improvement. The platform described above is applied at a UK-based company providing system solutions for manufacturing equipment to the customers worldwide. The scenario is the business of providing a complete air compressor system solution to customers, and then supporting this air compressor system throughout its life cycle by the after sales service. The air compressor business involves significant interactions with suppliers and customers and a real need to innovate and enhance this involvement.

There are two main user scenarios – see Fig. 5.12:

1. Collection of ideas when developing new air compressor system
2. Problem solving in breakdowns of the installed air compressor

The business involves working with suppliers to put together a compressed air system for spray painting. This means getting an air compressor, spray guns, nozzles, paint specification, and all the other parts of the system together. Then they are assembled together and installed at the customers' site as a complete solution. As with any installation where large volumes of liquid are involved in a complex engineered solution, there will be problems from time to time. Therefore the customer needs a great deal of support for any problems which arise, and this is the other part of this air compressor business.

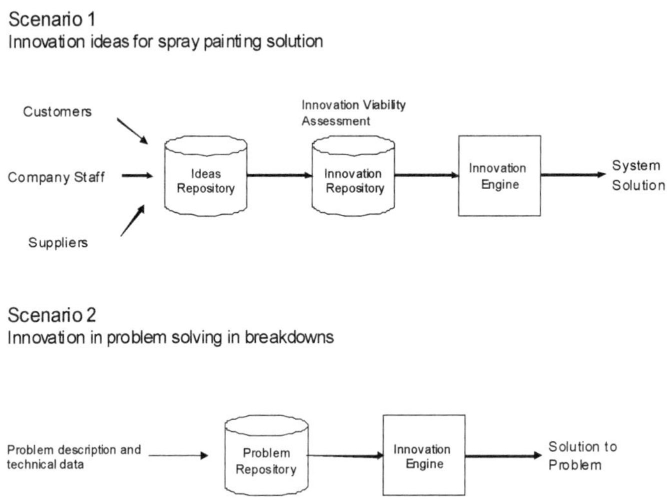

Fig. 5.12 Two scenarios for implementation of the platform within product innovation/improvements

The main goal is to develop a knowledge-based approach to achieve innovative solutions by collecting and using ideas from customers, suppliers, and employees and needs from customers and market and to turn these ideas cost-effectively into automation systems design solutions. For this, the company applies a system open to customers, employees, suppliers, and other partners from the network. The goal is the collection of ideas on needs that can be created by the novel technological solution. Considering the complexity and diversity of products in the network, the challenge is how to present potential innovations to customers and employees and motivate people to provide ideas on how to better use such technology (see Fig. 5.13).

The CAI is used by a number of experts within the company and a number of customers and suppliers are also involved and information provided by them was introduced into the repository. The company model was defined including some air compressor systems developed by the company and provided to several customers. In addition, about 190 problems related to these systems were also introduced. With this basic information defined, it was possible to begin collecting problems regarding actual and future systems, using CAI Collection services. The process of solving these problems also includes the collection of ideas.

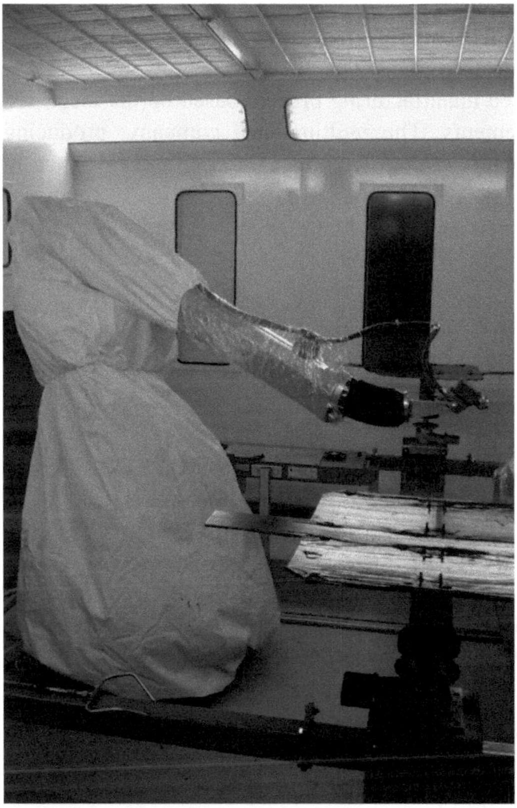

Fig. 5.13 Spray painting system

The main benefits of using the platform are:

- Developing and maintaining key customers through a close business to business approach, by providing them with interactive product information facilities
- Building the relationships with the suppliers, working together to improve the offer to customers
- Targeting new customers and transforming into new key customers
- Increased sales
- Reducing operating costs

- Improved staff motivation
- Reducing operating costs for key and targeted customers
- A more speedy approach in taking new products and services to the market
- Improve sales and customer support channel efficiency
- Global marketing
- The company becoming less vulnerable to any staff movements (absence or leaving the company)

Scenario 2: Innovation in multiple site manufacturing process. Scenario 2 addresses innovation in a manufacturing process. This scenario focuses upon innovation in a geographically distributed manufacturing process based on the identified problems and potential improvements. The multinational company, producing cans, has several plants in Germany, United Kingdom, France, and one each in Netherlands and Poland.

The CAI platform is implemented and used within several scenarios as depicted in Fig. 5.14.

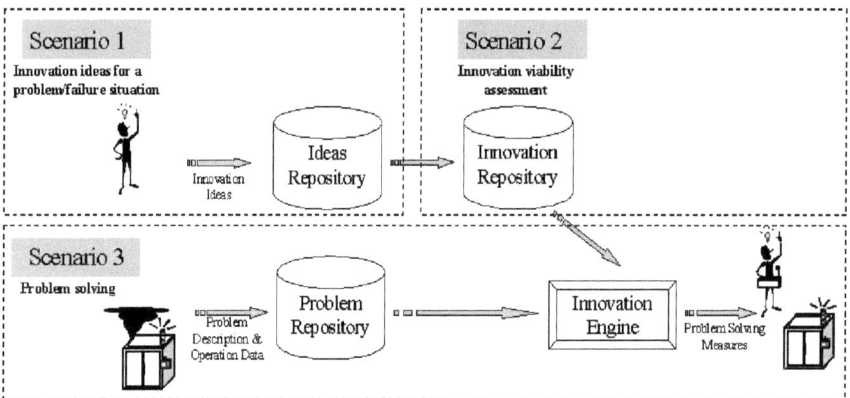

Fig. 5.14 Implementation scenarios within innovation of manufacturing process

The CAI platform has been implemented in several plants. In two plants in Germany it is specifically applied around the so-called Necker machine (see Fig. 5.15) which has the task of necking the cans and which was the critical bottleneck of both plants.

The platform is also applied in the manufacturing process of cans in two plants in Germany and Poland with the objective of sharing between both plants knowledge and experience on problems detected, overcoming cultural and linguistic barriers.

In both plants, the process where the CAI system is applied is a can producer transfer machine. This machine in the shop floor requires a generation/collection of innovative ideas in order to improve the process as well as the quality of the products. Since this bottleneck machines create a number of complex problem/failure causes asking for complex activities to remove these causes, CAI tools support a collection of ideas on these production steps. To solve the problems, this CAI system gives the staff members on the shop floor, as well as engineers, a consolidated overview of problems in production and proposals of causes/actions to remove these problems. The CAI system is first used by the process engineers, plant managers and selected maintenance workers who defined the appropriate plant model, especially the models of the bottleneck machines. This information was the basis for collecting problems and ideas.

Fig. 5.15 Necker machine in the shop floor

The services are used to collect ideas from different actors and over geographically distributed subsidiaries to solve complex process problems. The services provide information on problems identified in different forms depending on the expertise of actors, and gather ideas from shop-floor workers and process designers and support collaborative work on evaluation of these ideas. The services provide "similar" ideas in order to support ideas gathering and evaluation. The services are mainly applied in asynchronous collaboration, but their extensions to synchronous applications are ongoing. The challenging task is how to motivate shop floor workers to provide their ideas and collaborate on innovative solutions for process improvements.

The objective is to support cross regional teams building within different subsidiaries including international teams, where one of the key problem is appropriate (multi-language) ontology. The assessed benefits are:[10]

[10] These figures are based on the measurement of productivity and wastes before and after introducing the CAI services at a critical part of the manufacturing line and initial testing. About 350 problems and solutions registered on paper forms in both plants were introduced initially into the CAI system, using the collection services.

- Savings by application of the platform by shortening the time needed for collection and implementation of the new ideas by at least 30% (specifically reduction of time and efforts for solving product/process problems)
- Increase the number of innovative ideas on processes from employees by 50%
- Increase the number of innovative solutions of the identified problems within processes by at least 30 %
- (Indirect) improvement of process efficiency by 5% and reduction of spoilage by 7% (but once the system is introduced in at least eight plants world wide, the benefits, due to exchange of ideas on common problems, may bring improvements of over 12%)

On top of that, as a "soft" benefit it is expected to improve considerably the motivation of employees to contribute directly to the process innovations and strengthening of the company culture.

5.4 Collaborative Innovation Management in SME

This section is dedicated to ICT support to collaborative innovation management in small and medium sized enterprises (SME) driven Extended Enterprise (EE)

In general an integrated approach of organizational and technical measures is required to achieve a reasonable product and process innovation in any EE and especially in SME driven EE. Practice shows that also in SME the introduction of information technologies is the key technical measures in this context for innovation process.

In order to increase their competitiveness and revenue generation SME must look beyond the traditional design and engineering communities within the company to all resources, both internal and external, to foster innovation and improved design. This way, they must also use new tools and processes to involve customers themselves and other partners earlier and more directly in the product/process definition phase, and supply chain partners and other members of their value chain earlier in the development process.

On the other hand, SME must continue focusing on bottom line results by seeking techniques and tools that will allow them to control design processes, engineering, manufacturing, and servicing costs, without compromising product quality or customer satisfaction. Designing/innovating it right the first time as well as compressing timelines, reducing cycles, minimizing time-to-market and time-to-profit, and minimizing re-work have become desired goals of all product development processes.

SME need ICT solutions (services) tailored to their specific needs. Therefore, the solutions presented above have to be adjusted/extended to meet the SME needs. Three typical solutions for innovation in SME driven EE are described.

5.4.1 *ICT Services to Support Collaborative Innovation Processes in SME*

The ICT platform for collaborative innovation management, presented in Sect. 5.3, can also be applied for SME-driven EE. However, SME need additionally specific ICT services to support directly certain innovation processes. Large companies often have powerful ICT solutions which may be integrated within the collaborative platform to support innovation process but in general, SME cannot afford having many different solutions. They need intelligent knowledge-based assistants for designers, extending the SME knowledge base out to the customers, field engineers and suppliers, so that knowledge can be developed or added to throughout EE, promoting a virtual design and innovation process and creating the adequate environment for sharing product knowledge/information which includes:

- Effective feedback analysis
- Documents control
- Easy access to relevant information
- Design process planning and management

The *computer aided innovation* (CAI) platform presented in Sect. 5.3 allows inclusion of different additional services specifically needed by SME.

In the text to follow several such ICT services applicable for SME driven EE are described (Gorostiza *et al.* 2005; ASSIST 2006).

5.4.1.1 ICT Services

1. Feedback services. As many as 80-90% of product improvements/innovations come from users, therefore, from customers. If customers are satisfied, they give the company far more than mere payment of the product's price:

- First of all, they give loyalty. Satisfied customers will continue buying products from the same company
- Second, satisfied customers provide word-of-mouth advertising through the pride in using a particular product and through recommendations to families, friends, and colleagues
- Third, they supply feedback, or information about product improvement, customer satisfaction, and future desires and needs

These three elements – loyalty, advertising and feedback – are essential to a company's long-term survival. As has already been mentioned (see Sect. 1.1), dissatisfied customers, on the other hand, can bring a company to ruin.

A solution to facilitate such customer involvement is a set of ICT services (so-called feedback services) which support information exchange and collection, coming from customers but also from suppliers, from remote sites or different offices within the companies, as well as from other repositories or databases. The services also support a complete analysis of this feedback in order to use the output of the analysis as a basis for product innovation and new product concepts generation. The solutions for SME must be simpler than those presented in Sects. 5.2 and 5.3 (*e.g.,* on problems solving). Two subsets can be identified:

- Services for collection of feedback
- Services for analysis of collected information

The basis requirement defined for the feedback collection and analysis services is the large number of (potential) users who must be able to access the service. Therefore, this fact conditions its underlying architecture (see Sect. 5.3) which, in order to benefit from this source of information directly from customers or suppliers and SME's own staff, has to:

- Be open, so that information from as many people as possible can be collected
- Supply services to manage the information, both supplied to and received from customers
- Collect feedback from customers not only on problems, but motivate customers to provide recommendations/ideas
- Allow integration in the internal systems of the SME, in order to process incoming information, such as
  - Requests for information
  - Petitions from customers
  - Support to problem solving
  - Supply statistics

1a. *Collection of customer/internal feedback.* This is a set of SW services to support SME in the collection of the internal and customer's feedback for further processing and in the monitoring of projects and participation within EE.

For SME, sharing and exchange of feedback and potential improvement information across individuals and organizations can result in competitive benefits and added value in the improvement of existing products or activities or in the creation of new products or services, helping them to analyze how to achieve their most competitive advantage or to redefine their competitive position.

In order to process and introduce this feedback within the ICT platform, it is specifically needed to provide the adequate interfaces for introduction of this information coming from different sources in the innovation repository.

Depending on the criticality and privacy of the information, different communication channels are used for interaction between companies and actors. Normally, generic consults are done by telephone, while demand confirmation or formal communications can be done by fax, and exchange of drawings or specification documents by e-mail.

As explained before, considering that the Internet is a huge information and communication channel for SME, the acquisition means is provided through the

Internet. This approach is built upon so-called forum services. Forums or discussion boards can be considered as systems that gather the advantages of e-mail and chat or telephone systems for product support since they allow an ordered pattern communication channel and provide quick response and high interaction capabilities among the users which can be considered almost real-time.

Forum users are able to communicate each other in a very efficient manner. It is a knowledge-based system where all the information is stored for reuse. Users can upload images or files in order to support descriptions of problems or ideas for the improvements of new designs. The forum services are intended to improve the competitiveness of the SME by enabling the communication among the different actors of EE. In such a way, the designer is able to know the clients' needs, other designers' thinking, or the suppliers' opinions only by visiting the forum. Therefore, it is used as a tool for collection of the internal feedback and customer/supplier feedback.

1b. *Analysis and use of collected feedback/ information.* Once information from customers or potential customers has been collected and introduced in the system, SME must carefully analyze it, looking for customers' decision making processes and the customers' needs, preferences, and priorities. Problems, cases, consults... must be analyzed (customer's needs, product design specifications, problems with products, improvements requests) as feedback for the design department. This will constitute a basis for product/process innovation.

The analysis services have facilities to monitor and measure customers' needs and priorities, results from product design projects and assessment of quality and perceived value of products and services. SW services are able to assess database (innovation repository) and perform statistical calculations that will be presented in an appropriate form to the user. These services allow:

- Quantitative and qualitative measurement of feedback from customers
- Identification and analysis of strong points as well as weaknesses in products or services provided
- Analysis of external and internal improvements proposed
- Analysis of the developing process performance achieved and the results of the resources allocated and planning done by the company

2. Design manager. The design manager set of SW services are used by almost all people that are involved in the innovation process.

This set of services has to support:

1. *Collection of product/process data related design knowledge.* Using these services, companies can develop and maintain a database of structured design knowledge. They add or edit information related to a product design or information that is useful when carrying out design projects. The information is classified into categories/sub-categories. Metadata (like keywords and type) are defined for the information and used latter on as search criteria.

2. *Search for previous designs.* Users can search for previous designs that could serve as a basis for a new product design. After selecting an appropriate one from the result set, a copy of the existing design will be stored that can be then elaborated to a new product design.
3. *Access to the repository.* By using hierarchical or text search:
   - Hierarchical browsing of the information contained in the repository using a graphical user interface with a tree representing the information structure
   - Text search, with the combined use of multiple criteria (keywords, free text, *etc.*)

4. *Management of innovation projects.* The innovation manager provides comprehensive support to the users to:
   - Create new projects
   - Attach information to projects
   - Define the workflow for each type of projects
   - Create activities for a project and assign them to responsible persons
   - Define priorities for the activities
   - Manage the project activities and attach information to them

All the information generated during a design project is classified, stored and become available through the repository.

5.4.1.2 Application Scenarios

The CAI platform described in Sect. 5.3 and additional SME-driven services described above can be used in different ways in SME to support collaborative innovation processes. The platform has been applied in several SME within the research project ASSIST (ASSIST 2006). In the text to follow, five such application scenarios in real industrial environments are described: the first one in more detail and the other four roughly aiming to provide an overview of possible ways to use the platform in SME.

Scenario 1: Innovation process in SME producing customized products. This scenario addresses the application of the CAI platform and services in a German SME producing armoured and special cars. The platform is intended to be used for the re-design of individual product parts or the design of new product parts aiming to improve these parts based on the experience gathered during the manufacturing/assembly of current similar parts, as well as through the feedback from customers. The main objective of using the platform is to have a knowledge base where different type of information can be stored. It is very important for the company to store knowledge that is bound to employees, so that every staff member has access to the information even if an expert for a special task is absent or leaves the company. Another objective is to have a place where annotations or ideas regarding a car model can be stored. If any staff member has a good idea or gets feedback from a customer, the ICT platform is a good place to store that information (related to the car model in question) so that it can be assessed in the future. To find quickly a solution for a particular problem, it is important to store information on specific expertise of each worker.

An important aspect is the reuse of knowledge from former projects. When starting a new design project, the platform allows "copying" of the collected information from a former project and using it as a kind of framework or template. This is especially helpful when, *e.g.,* a customer asks for a new car that is similar to a model built in the past. The old product's data can be copied and modifications can be done with respect to the current requirements from the customer and to feedback that was collected earlier and stored in the knowledge base. This does not include the re-use of the data from the same car model only. Other car models from the past that are stored in the knowledge base could also contain relevant information that could be reused. Here CAI helps through its different search functionalities to easily find the desired information, *e.g.,* a solution for a specific problem that occurred during manufacturing/assembly of a car.

In addition to the knowledge base infrastructure, the CAI solution provides a simple framework for the project management as it allows planning of tasks within the innovation process to be done with due dates and the responsible employees. This is sufficient for small projects so that every employee involved can see which steps still have to be taken (or which have already been taken). Besides, the CAI platform also serves as a communication means between staff members, *i.e.,* a shop floor worker can insert a problem notice for the design engineers that they can analyze asynchronously and search for a solution.

The scenario describes a typical situation where CAI allows or eases the re-use of knowledge from former projects. A customer asks for a new kind of customized car similar to a model that was built for him some time ago. The knowledge from the former model should be reused for the new one which is a typical task for the CAI platform. The first step that has to be taken now is collecting knowledge about the old model from the database to allow its reuse. This is supported by the searching functionalities, where in the described situation search criteria would typically comprise the customer, the product family and some describing (key-)words or the product name. The selection of the customer company and the product family are shown in Fig. 5.16.

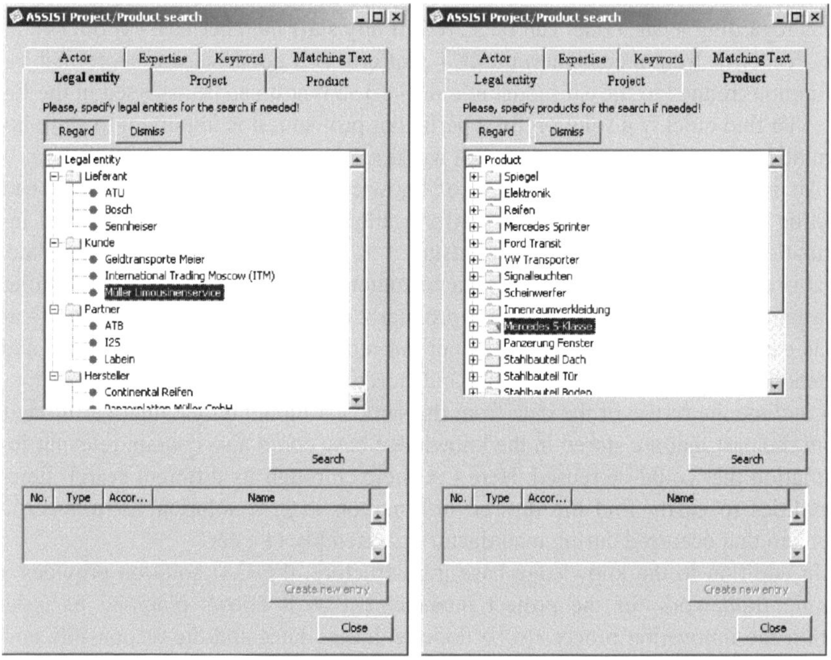

Fig. 5.16 Selection of a customer company and a product family as search criteria (ASSIST 2006)

After the former product has been found, a copy is created and given a new name, which can be used as a template for the new design to be developed. All related information is also copied and can be modified. After the new product entry was created a new project entry is entered into the CAI platform which contains a short description of the design project, data about the customer who ordered the new car model, and a relation to the product copy created before.

Then the planned activities for designing the new car such as meetings of the design team, talks with the customer, parts manufacturing and assembly, as well as preparation of specification and drawings are added to the system together with planned dates and responsible employees (an example of a new activity is shown in Fig. 5.17). Through these functionalities the CAI platform provides a simple framework for the management of the innovation project.

The actual design work is done collaboratively by several employees who, besides product designers, could also be shop floor workers who may have suggestions regarding the design to ease manufacturing and assembly of the new car model. The customer may also contribute by refining the specification or by giving feedback to intermediate project results.

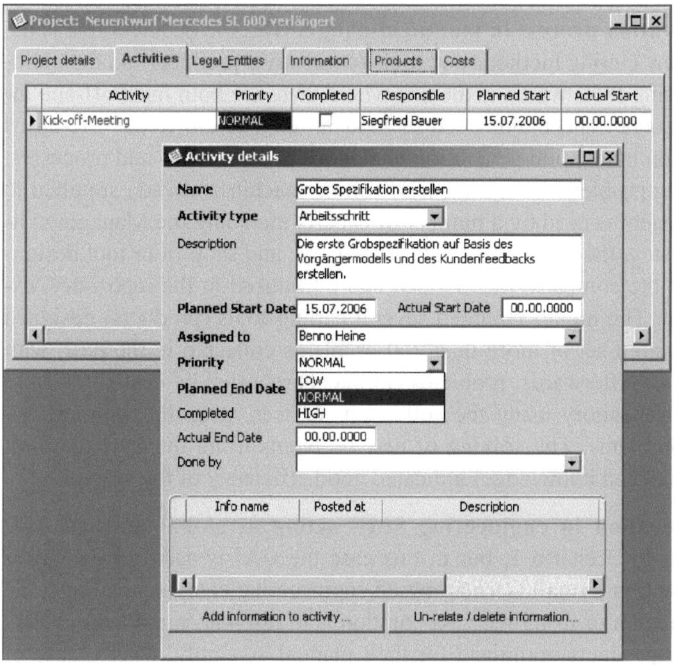

Fig. 5.17 Adding a new activity to the innovation project description (ASSIST 2006)

Reuse of product components used in similar car models is also possible and supported by the CAI platform through the functionality to store the hierarchical structure of a product. After the design phase is finished the next step can start: the manufacturing and procurement of product components and the assembly of a prototype of the new car model. If problems occur during this phase (*e.g.,* some parts do not fit exactly during assembly) they are inserted into the system by a shop floor worker as feedback for the product designers. The worker has to choose the product and add an information object (*e.g.,* a textual description) to it which is assigned to the category "problem."

The design team can (asynchronously) read out the problem report and use the CAI platform to find similar problems and their solutions from former projects. After studying problem reports and solution descriptions (and perhaps a consultation with the reporting shop floor worker), the designers make some changes in the design to prevent new appearance of the reported problem. A description of the chosen solution is stored in the repository related to the problem report so that it can be reused in the future. Thus the CAI platform helps to avoid some mistakes in early design phases that otherwise would lead to problems in assembly or use of the new car.

The measurements of the benefits of the application of such solution indicate savings of more than 30% in time and efforts needed for development within design projects.

Scenario 2: Innovation process in manufacturing SME. The business line is the development of new cutting methods and tools (company's product) and the maintenance of the cutting tools which is the one where there are both most off-site involvement of personnel and major need of innovation. Therefore, the UK-based company focuses on innovation generation to achieve new products and processes, and also on solving problems occurring with cutting machines already supplied to customers. The system is used by a number of experts including the Managing Director of the company, the Works and Sales Managers, and shop floor tool designers. The model of the company was defined and introduced in the repository, using the set-up tool. The model included several cutting tools (products) designed by the company, and a set of more than 200 problems collected in the past, with respective solutions. Afterwards, problems/comments reported by customers were introduced in the repository using the collection services, with the complete description of the problems. The solving of new problems using the CAI platform (based on the introduced knowledge) indicated good efficiency of the system.

Scenario 3: Innovation in engineering SME acting at global market. This scenario is similar to Scenario 1, but in this case the CAI is used for design of various products within a small German-based engineeering company acting at the global market. The aim of using the CAI solution is to provide support for project teams redesigning products optimized for their manual assembly, where reuse of knowledge from old product designs is essential. The services help through the search functionalities to find easily the appropriate knowledge and allow an easy reuse by providing functionalities to duplicate data from former projects. The innovation repository allows capture of information from many different projects that deal with a variety of products from different areas. It is important to save the information in a database as the participants change between projects which are carried out for different customer companies. The services presented allow the knowledge to be stored and reused independent from particular people. Besides information about old product designs, the storage of a project's activities is supported as well, which could help start a new project, serving as a proposal of tasks to be done.

The main aspect is comprehensive search functionalities to find products and solutions for design problems similar to the current ones. This is especially important because of the variety of products from different areas that are redesigned. It is essential that information in a large knowledge base can be found fast and easily. When a product has to be redesigned, the team wants to reuse former project and/or product information that best fits to the current project. Different search criteria can be used such as product families, project types, companies (supplying product components), keywords, and matching text. The services return products and projects that match the defined search criteria and this also allows creating a copy of an appropriate search result to be used as a template for the new design. All duplicated information can be modified. Individual parts or components can be added, removed, or replaced by others which are, *e.g.,* easier to assemble or have a lower price.

Information from other designs than the one copied could also help in redesigning the product. This could be knowledge from many different products that were redesigned in the past, mainly generic concepts or patterns for the solution of particular tasks that occur when products are optimized for their manual assembly.

Scenario 4: Training staff in SME. The scenario addresses application of the CAI platform for training purposes in a Spanish SME, developing and commercializing a new line of products for the nautical, underwater and leisure sector. These products are named as telepresence that means a remote interaction in an underwater environment.

The usage of the CAI services helps in the process of "conceptualization." It allows gathering external information and attaching it to the product requested by the customer. Moreover, it allows for incorporation of the "voice of the customer" into the knowledge base, so one can get that information when needed and in a natural manner while the designing and manufacturing process is running.

It also allows one of the most important features for SME: increase the intellectual capital of the company and efficiently train people (*e.g.,* train new engineers on innovation issues). The SW services can capture the experiences of the junior engineers who enter the company for a short period of time (job practices in order to improve their professional skills). As SME has an important rotation of employees, the tool allows a capturing of this knowledge. New engineers can learn more quickly as the know-how of the previous tasks is stored inside the system. This improves productivity and widens the portfolio of products.

In addition, the CAI platform improves the reaction towards clients and potential clients. All problems with clients can be stored and correlated to the products, so when the client calls back, the staff members already have his complaints or suggestions in mind. Besides, when a client asks for a partially modified product, it is not necessary to "start a project from zero"; instead, it is possible to adapt one existing project related to that particular product and have access to all needed information on technical solutions available in the market or information on patents. It is possible to easily redefine the planning and the allocation of resources simply adapting a couple of tasks to fulfill the current user requests. This information retrieval at the right time permits to give the customer a quick response.

Summarizing, the SW services improve the productivity by:

- Increasing the interchange of information among different departments of the organization
- Feeding the knowledge repository with internal source data, such as employees' know-how and feedback, product portfolio or common designs and with external source data, such as feedback from the customer, customer requirements, different assemblies and standards from providers or marketing and legal issues
- Planning of resources based on previous similar designs

- Providing the right information at the right moment (*i.e.,* when a product is designed, the tool is used to analyze information about technical solutions available in the market, patents, verification protocols, *etc.*)
- Having a manufacturing processes knowledge base with graphs and animations on collected problems for training junior engineers

The CAI system specifically exhibits high efficiency in training junior engineers. It is assessed that the training period for the young fellows reduces by more than 50%. The junior engineers can very flexibly use systems for training on innovation (in parallel to overtaking other tasks in the company).

Scenario 5: Innovation process in networked manufacturing SME. The scenario addresses the application of the CAI platform in a small company with less than 50 employees in the business of designing and manufacturing thermostats, networked with several subsidiaries geographically distributed (Spain, Turkey, China). The main business objective of application of the CAI system consists in adapting the thermostats to new requirements. These requirements can be gathered in the form of queries and comments. The storing and management of these information allows for their re-use in the design processes.

The CAI platform provides an inexpensive framework and means to obtain a structured storage of information. The platform allows an efficient movement of knowledge among the different areas of the organization and among geographically distributed subsidiaries. The employees can obtain the right information at the right time. The commercial department collects the requests and feedback from the clients by phone and e-mail. Then they insert the information in the knowledge base. This information is linked to a product or a project, so it turns into explicit knowledge. The commercial department answers to the client and the technical department provides a technical description to the client if needed. Furthermore, the technical department can attach any complementary information to that particular product or project the client requires. The knowledge is always available and the problems encountered, elements or documents related to a particular project or product are stored conveniently in the system.

The platform also facilitates internal communication, *e.g.,* between the production department and the technical department. This allows valuable feedback, which accelerates the design process. This issue is critical, for example, with the production plants distributed world wide (Turkey, China).

When a client asks for a new product, which is similar to a previous one, the employees can obtain benefit from the platform and, instead of starting from scratch, make a new project based on the previous one. This way the workflow is defined and required resources (drafts, calculations, planning schemes, legal information, quality plans, *etc.*) from the old product can be "re-used." Once this skeleton is build up, the user has to modify/add/delete the resources needed to accomplish the work for the client. This dramatically reduces the preparation phase and the technical and economical viability assessment of the new product development, so the client request can be responded in much shorter time frame. This leads to an operational improvement with clients. The average faster answer to the clients inspire them a high amount of confidence and increases the probability of closing more contracts.

Due to a high similarity of the products, the platform allows reduction of response time to a customer by more than 55% in average.

5.4.2 Combination of e-Business and e-Innovation Solutions for SME

The *computer aided innovation (CAI)* solution to be briefly discussed in this section combines different functionalities needed for *e-Business* and for *innovation.* As the innovation in modern SME requires strong networking among companies, there is a logical need to combine functionalities supporting *e-Business* networking and knowledge sharing within so-called *e-Innovation* approach, *i.e.,* innovation supported by modern Web technologies. The presented platform is specifically developed for SME in construction sectors since in this sector needs for such combination of e-Business and e-Innovation are especially emphasized.

Such a combination of business and knowledge community is likely to be one of the key future trends in innovation in the twenty-first century. Therefore, it will be further considered in Chap. 6 dedicated to future trends. In Sect. 6.5 a more advanced (agent-based) ICT solution for such a combination will be presented. The ICT solution presented in the text to follow, although on the edge of the state-of-the-art (based on semantic knowledge technology), is from implementation point of view simpler than the one to be presented in Chap. 6. From the business point of view the platform to be described supports less advanced forms of networking among SME than that in Chap. 6.

5.4.2.1 ICT Services

The CAI presented is partly developed and investigated within the research project Know-Construct (2007). The CAI is a common platform providing a combination of two general functionalities (Sorli *et al.* 2006b, 2007a, b):

1. *Customer Needs Management System (e-Business).* A decision making support system regarding the product characteristics, product applications and related consultancy services within the networked SME
2. *Knowledge Community Support System (e-Innovation).* A system to support an efficient formation of advanced communities of SME, through their specific knowledge integration, management and reuse *via* a common knowledge base

The system responds to the following aspects:

- Low cost for the involved SME and customers
- Internet-based
- Efficient customer support
- Support an open collaboration between actors in the business processes
- Record the key information associated with all the involved actors, assuring traceability
- Assure common terminology and ontology
- Assure high security
- Support mobile users (knowledge available in a form of essential expertise, reachable anywhere, at any time), *etc.*

The overall functional architecture of the system is represented in Fig. 5.18. The main functionalities of both *Customer Needs Management System* and *Knowledge Community Support System* modules are as follows:

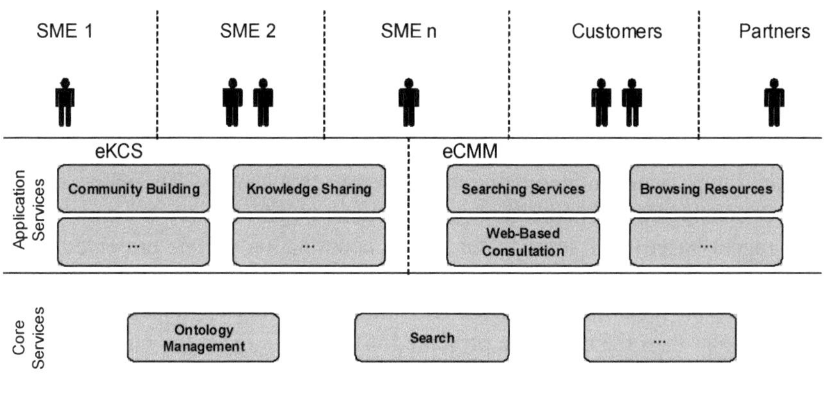

Fig. 5.18 Overall Functional Architecture of the system

e-CNM – Electronic Customer Needs Management. See right hand side of Fig. 5.18. e-CNM utilities are as follows:

- Searching (SME provided) services: facilities to search information and knowledge related to services provide by SME
- Browsing community resources: facilities to browse freely information about the companies belonging to the community: products, services, procedures, *etc.*
- General browsing: facilities to browse in a structured way through the information made available for the customers
- Searching materials/products/components/procedures: facilities to search information and knowledge related to materials
- Interactive, Web-based consultancy: tools to help customers to solve problems and get advice

e-CNM module also features a portal service, providing the customers with access to individual community members (SME) e-commerce/e-business systems, integrated with information search and consultancy functionalities. From the community members point of view, e-CNM provides customer relationship management functionalities in terms of collecting and organizing feedback and knowledge from customers, and managing consultancy services.

The e-CNM core services (see the text to follow) consist of the semantic resource management which represents the central knowledge processing and management services of the system.

e-KCS – Electronic Knowledge Community Support (left hand side in Fig. 5.18). The e-KCS module covers the following assets:

- Knowledge community building: tools to create and share knowledge through collaboration, like discussion forums, *twiki* tools,[11] news services, *etc.*
- Knowledge sharing: tools to collect, disseminate and search experiences, problems, best-practices, opinions within the community
- Content management: tools to classify, organize, search documents, *etc.*
- Knowledge structure management: tools to manage ontologies and classifications schemes
- Information collector: collection and organization of information from external sites and portals
- External search manager: complement searches in the community knowledge with searches to external sites, portals, databases, *etc.*

Semantic Web technologies are fundamental for eKCS in order to provide information retrieval, both internally and externally to the knowledge community.

Similarly as in the case of ICT platforms presented in previous sections (*e.g.*, that presented in Fig. 5.2. or that presented in Fig. 5.7), this generic architecture can be broken down into two layers (Soares 2005):

- Application layer: application services and systems/applications layer including above listed functionality
- Core service layer: core services are divided in semantic resource management and a set of functionalities that provides the systems/applications with access to the semantic resources, namely: ontology manager, indexing and knowledge extraction, semantic searching and navigation, aggregator/integrator, business data model wrapper

---

[11] Wikis are typically used as shared whiteboards that allow users to add, remove, or otherwise edit all content very quickly and easily. The ease of interaction and operation makes a plain wiki an effective tool for collaborative writing and to share knowledge. Structured wikis provide database-like manipulation of fields stored on pages and usually offer an extraction and presentation language or markup. TWiki is a structured wiki, typically used to run a collaboration platform, knowledge or document management system, a knowledge base, or team portal. Users can create wiki applications using the TWiki Markup Language, and developers can extend its functionality with plug-ins.

System knowledge base. The system database is composed of the *business database* and the *semantic resources databases*. The former supports the eCNM part in answering to the requirements of decision support system functionalities and eKCS in provision of the community specific knowledge. The latter is divided into:

- A business metadata database to store the metadata descriptions of the information/knowledge of the knowledge community
- A semantic resources database which stores the industry knowledge high level ontology and the local ontologies

5.4.2.2 Applications

The CAI described above was applied in the construction sector. The construction sector is characterized by a high level of fragmentation, with a large number of participants in each construction project, the large majority being SME. To increase flexibility and profitability the bigger construction companies have significantly reduced the scope of their activity and consequently the number of employees, focusing on the core tasks of the construction process and subcontracting most of the work to specialized and smaller companies. A narrow technical specialization must be replaced by significantly wider technical competence through integrated teams such as knowledge communities, combined with on time, within budget works completion. In addition, SME need to improve communication with their customers in order to provide better product support and services. Necessity of knowledge and competence integration for a successful response to customers needs imposes a need for establishing the knowledge communities of SME.

The critical issue is the approach for knowledge representation and ontologies *i.e.,* to apply adequate domain related *ontology*, as well as a classification system for this sector applicable in SME environment.[12]

The benefits of application of the described CAI can be summarized as:

- Improved innovative technical support to product and service users (customers)
- Better access to wider technical competence required to satisfy customer needs, through closer co-operation and knowledge exchange among SME within knowledge communities
- Improved quality/price ratio
- Increased innovation capacity
- On time completion of increasingly complex tasks

The platform described can be used in several different modes according to users' desires and customizations. As examples, six scenarios have been tested along

[12] Solutions applied within another projects of the kind (e-COGNOS- http://www.e-cognos.org/) were partially re-used (Sorli *et al.* 2006).

the research project Know-Construct (Sorli *et al.* 2006). These are briefly presented in the text to follow.

Scenario 1: Search for product information/documents. Potential customer looks for information about a specific construction product. The user logs-on in *e-CNM system* and selects the product through the product categories/groups which has been constructed according to the ISO 12006-2/EPIC Table. After navigating through an appropriate number of steps (in each step an appropriate number of selection boxes/lines are offered) a set of products is presented. The user chooses one of the entries of the result set to get more detailed information. The product's/manufacturer's Web-site is displayed with detailed information, also providing links to documents with more specific information about the product, application notes, *etc.* A similar scenario can be applied when a potential customer looks for information about a specific construction work/service provider.

Scenario 2: Search for partner. SME user looks for a partner with a specific competence (complementary or identical to its own) in order to answer to the tender for a project or to innovate product/service which is beyond its own possibilities. The user logs on in *eCNM-system*, and enters "search for partner". The three categories of competence (1) products (2) works and (3) entities/objects are offered. Each one can be searched through as described above. The system also looks for the existing offers of other members of the community and sends (automatically and/or manually) to the selected potential partners an invitation/call to participate. Initial contacts are also possible through the system.

Scenario 3: Customer feedback. The system user (customer) wants to insert some specific feedback related to a product implementation or functionality or to an improvement of characteristics. In order to facilitate the insertion process and to facilitate the subsequent analysis of the inserted comments/suggestions and their structuring, the customer will be led through the areas described above up to the specific product, service or other topic. During the feedback insertion the customer can use some of the functionalities described above as support for precise feedback definition (see Sect. 5.4.1.1).

Scenario 4: Navigate legal information. A work's supervisor needs to have an idea of the legislation related to a given construction area. This person does not know exactly what and where to look for the information needed. The user then uses the semantic navigation facility of *e-KCS* that enables him to browse sets of legal documents and notes according to several categories and perspectives. An example of the interactions between the user and the system is shown in Table 5.7.

Scenario 5: Register problem/solution. A member of a company belonging to the community/SME network finds a solution to a given problem relevant for the community. The user looks in the *e-KCS system* to see if something similar is already available, possibly navigating through several problem categories. In the case of being a new solution, or a new problem, the user introduces the information about it and the system assists him in finding the right category to classify it.

Table 5.7 Interaction user/system

User	System
Selects "Navigation" from the system options	Presents a top level graphical view of the classification scheme
Selects "Legislation" from the top level categories	Presents the next levels of the classification (legislation subcategories)
Continues to go down the hierarchy until finding a category fitting with his/her interest	
Selects the option to show the content classified in the legal category previously selected	Shows a list of content

Scenario 6: Consolidate product's application experiences. A community member needs to gain some insight regarding possible problems in the application of a given construction product. For that, it is possible to search for feedback from customers from all the companies belonging to the community. This way, some recurring application patterns can be found and correlated to the problems reported. This is an example of how individual information (previously collected) on customers' needs *(e-CNM)* can be shared in the Knowledge Community *(e-KCS)* and be used by any community member.

5.4.3 Collaborative Knowledge-based Engineering Solution for SME

As explained above, collaboration is particularly vital for product design since this upstream activity in the product life cycle has a decisive impact on the success of the particular product (Horváth *et al.* 2002; Chaudhari and Patil 2002).

In addition, it is becoming obvious that it is not possible to fulfill the new requirements solely based on conventional *computer aided design (CAD) – computer aided manufacturing (CAM)* systems and the current Internet facilities (Sorli and Gutiérrez 2002). As new infrastructure, tools, methods, and knowledge are needed, a distributed cooperative product design capability is necessary.

Generic use case: knowledge-based distributed product design and manufacturing system. This *generic use case* manages the distributed design and manufacturing process among different teams in SME distributed over the world by the Web, including managing the relevant knowledge for the design and manufacturing processes.

The approach presented, developed within the WECIDM project[13] (Mendikoa *et al.* 2005; Mendikoa and Sorli 2005; Sorli *et al.* 2005), mainly focuses on those techniques that can support multi-distributed clients and provide a dynamic database service, thus making possible a dynamic distributed design and manufacturing process. The core application of the system manages the distributed design and manufacturing process between different teams through the Internet, including the management of all the relevant product knowledge for design and manufacturing processes.

Figure 5.19 shows the block structure of the distributed product design and manufacturing system whose modules are described in the following text.

As can be seen in Fig. 5.19, both *designer* teams (CAD users) and *manufacturing engineers* (CAM user) are connected to a common middleware (left side of the figure) thus being enabled to interact on line from distant physical locations.

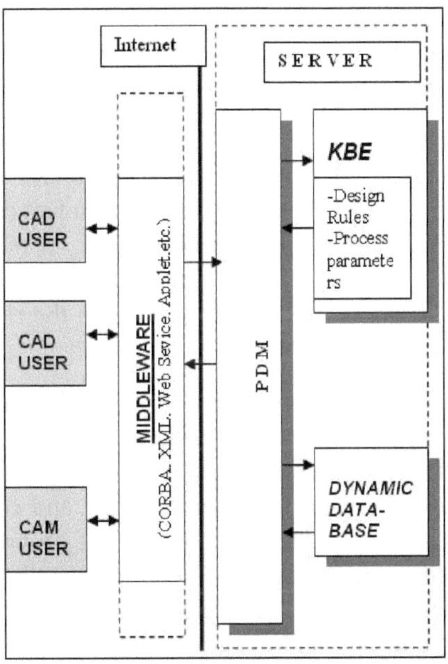

Fig. 5.19 Structure of the distributed product design and manufacturing system

Using the specific case of forging part design and manufacturing as an example, the distributed design and manufacturing methodology through the tool developed would be as follows.

Manufacturer engineers introduce or modify the design rules parameters. Designers will be able to get automatically a design in their local CAD system incorporating the design rules selecting the specific product family, part dimensions,

[13] WECIDM (2004) Web-enabled Collaboration in Intelligence Design and Manufacture. (Asia IT&C ASI/B7-301/3152-99/72553)

and process alternatives. This information (family type, part dimensions and process alternatives) will be stored in a file residing in the central server that the designer can download. This file will contain as well the current design rules fixed by the manufacturer engineer, and, in this way the designer can automatically incorporate these manufacturing rules in the design.

For the designers (CAD users) to be able to apply the process and geometric information automatically in the CAD package, a two-dimensional (2D) CAD parametric "template" must have been developed for that CAD package and that particular family. In this example the parametric template developed corresponds to a specific forging family, in this case corresponding to a rotational part.

This template resides in the designer local system. The CAD model is thus generated applying the information introduced by remote users, and therefore it fulfils the forging design rules required by the manufacturer.

Neutral format is used and uploaded (to the remote central server) by the designer user, so that the CAM user can get the geometry of the part and is then able to generate the CAM files from the geometry and launch automatically the manufacturing process onto the machines driven by computer numeric control (CNC).

Description of the system. The basic structure of this system for distributed product design and manufacturing to be described in the following text follows the structure shown in Fig. 5.19. The system includes dynamic database, product data management (PDM) and Knowledge-based engineering (KBE) modules. As mentioned before, designers (CAD users) and manufacturing engineers (CAM users) interact with the server through the middleware. The central server (on the right side of the figure) includes:

- *Product Data Management (PDM).* The PDM application performs the basic data management features and manages the knowledge-based engineering (KBE) and the dynamic database (DD) modules (see below). Modifications in the files and databases in the *central server* are done hierarchically and controlled by this PDM application. An assembly may be composed of different sub-assemblies, each in turn composed of different parts. Every part has different files associated, corresponding to geometry, CAM files, and any other file containing information relevant to the design and manufacturing process for that part. The PDM application is linked to a database where all the relevant information related to the assemblies, parts and documents is stored. This database in not visible to the user, whose only interaction with it is through the PDM tool.
- *Dynamic database (DD).* A central dynamic database stores the product/process "knowledge" through its value chain. The DD interacts with the PDM under control by a local application. Both modules perform the main product data management features.
- *Knowledge-based engineering (KBE) module.* Specific module centralizing the design and manufacturing process. KBE modules for different kinds of parts and production processes (such as forging, machining, *etc.*) are available since

different processes have different type of rules related to them. The set of rules includes the necessary knowledge for the design and manufacture of the part. KBE allows companies to capture and reuse the knowledge and experience of their engineers, together with manufacturing best practices, legislation, costing, and other rules for the product development. Different modules may be developed for each process and for each part family and plugged in, in order to implement the specific design rules and process parameters. These modules are implemented in connection with the dynamic database where the design rules parameters values are stored.

Designers can in this way get the values for the different parameters in order to apply the adequate design rules in the product design. These data will be automatically used inside the CAD system through the appropriate Application Programming Interfaces (API) developed for that specific CAD package. The users select the remote KBE module and with the correspondent API will be able to work within the local CAD package.

Through these three modules, the *server* contains all the project information, *i.e.,* every file related to the product (geometry, process parameters, *etc.*) and external users can interact with it through this specific PDM application.

From the users' point of view, their respective CAD or CAM system should include a specific Graphical User Interface (GUI) through which relevant data can be introduced and visualized. The middleware includes the necessary tools in order to ensure the correct communication and visualization of data.

Figure 5.20 shows an example of graphical user interface (GUI) that allows either the *manufacturing engineer* or *designer* to read or write the values of some design rules parameters corresponding to a typical forging process, *i.e.,* parameters such as flash land geometry, pre-form volume, draft angles, convex radius, *etc.*

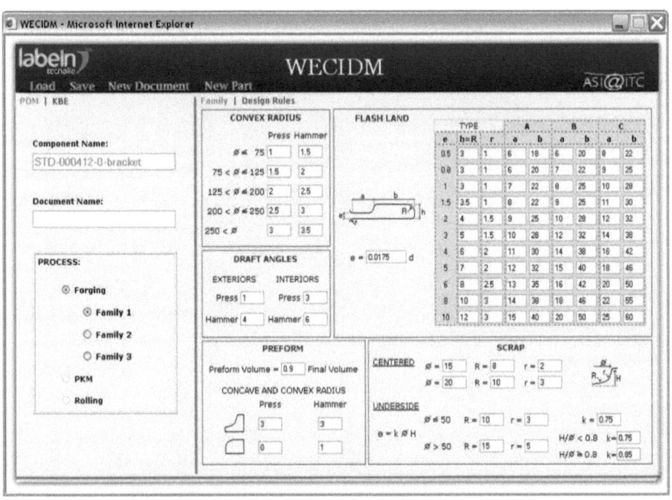

Fig. 5.20 Design rules for a forging process

Chapter 6
Future Trends

Abstract The product/process innovation in industry is undergoing change. The approaches to support innovation including ICT tools have experienced tremendous changes in the last few decades and are likely to be a subject of further evolution. This chapter provides an overview of the key trends in product/process innovation in industry. The emphasis is on new approaches for innovation which are already emerging and they will find wide application in industry. Especial emphasis is made upon ICT as a key facilitator of the innovation process in future. Many research trends could be observed which are likely to provide new innovation approaches and effective means to support such new innovation processes. This chapter focuses upon those approaches which are characteristic for future trends and which are at least partly already applied in industrial practice or are likely to be applied in the next few decades. The chapter includes conceptual trends in design and innovation processes (eco-innovative design, lean design, open innovation) and technological trends (ICT solutions for innovation in non-hierarchical networks, collaborative working environments for collaborative innovation with emphasis on semantics aspects), as well a new methodological approach in design (axiomatic design).

6.1 Introduction

To some extent, the book as a whole can be considered as "future trends." Mainly what has been presented and discussed in Chaps. 3 through 5 is, to a certain extent, ahead of the state-of-the-art; most of the achievements are prototypes that are still in their childhood period but are expected to be incorporated shortly in many industries all over the world. So these are in fact "future trends."

However, this chapter is aiming to discuss and present several new concepts, methods, and ICT solutions much related to the overall content of the book which nevertheless, have not been introduced in more detail since they can be considered still further ahead in time and (unlike the others) have not yet been sufficiently explored in industrial practice.

Within this approach, the present chapter has been structured as follows:

- Conceptual trends (Sects. 6.2 – 6.4)
- Emerging tools (Sects. 6.5 – 6.7)
- New design methodology (Sect. 6.8)

6.2 Eco-innovative Design

The way products are designed has an utter importance on their environmental impact along their life cycle: how they are produced, used, and disposed of at the end of their life.

The concept of "Sustainable Consumption and Production" (SCP) is becoming quite popular recently (NATO 2007; Charkiewicz *et al.* 2001; Cohen and Murphy 2001; OECD[1] 1997) raising awareness of the close relationship among production, consuming habits and environmental footprint. Above all, stands competitiveness (Coenen and Díaz López 2008) as the driving force that moves industry ahead. Finally, innovation is a must to combine the other two parameters (sustainability and competitiveness) as shown in Fig. 6.1.

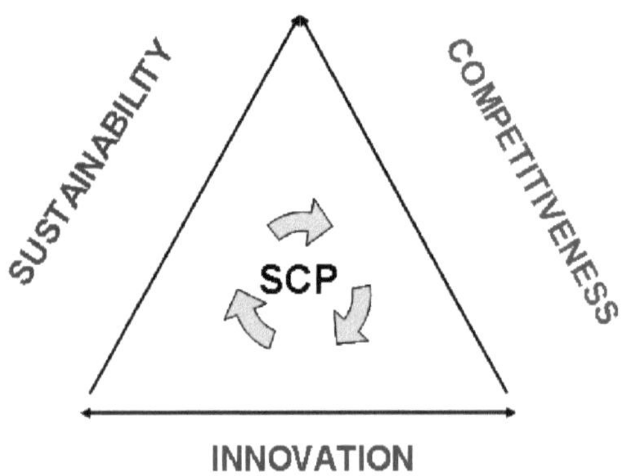

Fig. 6.1 Sustainable consumption and production

Consumption greatly depends on the societal mentality which can be educated and conducted by the governing bodies both positively through incentivizing and negatively by regulations and penalties, but also in the way the products are delivered. The concept proposed by McDonough and Braungart in their book "From

[1] OECD: Organisation for Economic Co-operation and Development. http://www.oecd.org/home/

Cradle to Cradle" (McDonough and Braungart 2002) of *product of service*[2] goes in that direction.

The gradual introduction of eco-friendly techniques in the manufacturing industry is an urgent necessity faced with the alarming depletion of the natural resources and the growth of the environmental impact generated so much by products throughout their life as well as by the manufacturing processes necessary to produce them (Sorli *et al.* 2002b).

As early as 1998, the European Commission started commissioning some studies (Ernst & Young and SPRU 1998) consulting experts and stakeholders on the issue of the *Integrated Product Policy* (IPP):[3]

> All products cause environmental degradation in some way, whether from their manufacturing, use or disposal. Integrated Product Policy (IPP) seeks to minimize these by looking at all phases of a products' life-cycle and taking action where it is most effective.
>
> Ernst & Young and SPRU

to finalize with the publication in February 2001 of a "Green Paper on IPP" (European Commission 2001) in which it is said that:

> This Green Paper proposes a strategy to strengthen and refocus product-related environmental policies to promote the development of a market for greener products.
> The strategy is based on the Integrated Product Policy approach and intends to complement existing environmental policies by using so far untapped potential to improve a broad range of products and services throughout their life cycle…
>
> European Commission

In current "Competitiveness and Innovation Framework Program 2007–2013"[4] (CIP), the European Commission integrates the Eco-Innovation Program" in which the concept of eco-innovation[5] is defined just enhancing and adding the "eco-flavour" to the 1995 definition (see Sect. 2.1):

> Eco-innovation: All innovation, which can benefit the environment.
> Eco-innovation is a fairly recent business & technology area which may be described as the production, assimilation or exploitation of a novelty in products, production processes, services or in management and business methods, which aims, throughout its life cycle, to prevent or substantially reduce environmental risk, pollution and other negative impacts of resources use (including energy use).
>
> European Commission

---

[2] The product of service intends to fulfill the customer expected requirements within the "service" concept: *i.e.*, the customer does not buy an automobile but hires a transportation service.

[3] http://ec.europa.eu/environment/ipp/home.htm

[4] http://ec.europa.eu/cip/index_en.htm

[5] http://ec.europa.eu/environment/etap/ecoinnovation/def_en.htm

New tendencies of eco-innovation raise a radical shift of paradigm since "less pollution or reducing the environmental impact" isn't sufficient anymore[6] and it is necessary to shift to "products and productive means positive to the environment."

This can only be obtained from the concept of "eco-innovative design." One of the examples that are usually mentioned in this aspect is the integral utilization American Indians made out of the buffalo in which as much as practically 100% of the "product" contributed benefits to the users and to the environment, maintaining an ecological balance over hundreds of years (Braungart and McDonough 1998; Zlotin *et al.* 2002).

This objective that at the current moment can seem Utopian should be the practical approach allowing us to move towards the "*ideality*" concept for the products and their manufacturing processes proposed by TRIZ methodology (See Sect. 1.4.4). *Ideality* has to be understood as: "To provide the required functions minimizing any negative effects." Among negative effects, usually the most important are the cost and the environmental impact.

Three different approaches can be considered as can be seen in Table 6.1.

Table 6.1 Eco-design

Intervention	Comments	Environmental improvement
Improvement in current products	Products re-design must take into account environmental aspects balancing several aspects as: phase in the life cycle ("S-curve"), cost, expected environmental benefit, *etc.*	Low. Improvements in living existing products can't be high. In general what is not considered in the original design is very difficult to solve later on
Improvement in current manufacturing processes	Being life expectance of the processes higher than in products, some environmental positive improvements can and must be implemented	Medium. Processes are very much anchored by their original design but some improvements may be reached in terms of reduction in energy consumption, reducing waste, *etc.*
Eco-innovative design of new products and processes	Incorporating from the conceptual design all possible environmental positive benefits is the future road to be tackled	Very high Shifting paradigm towards "ideal" (TRIZ approach) products and processes conceived to be positive to the environment

Therefore, product innovations must follow the "from cradle to cradle" concept (McDonough and Braungart 2002) taking into account that the ecological ap-

[6] According to the Global Footprint Network: "In 2005, humanity's footprint exceeded the earth's biological capacity by over 30 percent"
http://www.footprintnetwork.org/en/index.php/GFN/page/methodology/

proach is only feasible if it is considered just from the design phase of the product and its productive processes.

In the above-mentioned book authors make a distinction between "eco-efficiency" as "doing more with less and being less bad" and "eco-effectiveness" as "working on the right thing." The concept they discuss is that by the current circumstances, it is not valid anymore being eco-efficient trying to reduce waste and environmental impact but what is really needed is to be eco-effective, imitating nature and eliminating even the same concept of waste.

Eliminating the concept of waste implies considering any product as part of a whole close metabolic system in which end of life of a specific component means nutrients for a new one. In that sense, they identify two cycles that should never be mixed up:

- *Biological metabolism* or biological cycles in which the end of the product cycle can safely go to the earth and biodegrade
- *Technical metabolism* in which the material or products are designed to come back to the technological cycle being "up-cycled," meaning that recycled products do not lose any characteristics or properties (down-cycling) but can actually be used as brand raw material for any purpose: up-cycling

The proposed concepts and some of the ideas are a clear trend that has to be followed in the near future. Of course what is still missing is *how* this new revolution has to be started (maybe it has already) and conducted.

On the other hand, it is difficult in practice to separate "*efficiency*" and "*effectiveness*" since both must be tackled in conjunction: the need is to produce the right things (effectiveness) at minimal costs and efforts (efficiency). Combining both concepts can be considered "*eco-innovative design.*"

In some sense, this new revolution under the paradigm of eco-innovating design is following the "coming back to the handcraft times" cycle that has been analyzed in Chap. 1.

As McDonough and Braungart point out in their book, the first industrial revolution is in the origin of today's environmental problems that at the time were completely dismissed since the concept of resources depletion were absolutely out of consideration.

In contrast, coming back to the pre-industrial times means making use of local renewable resources: the American Indians sustainable exploitation of buffaloes, houses made of clay blocks that keep it warm in winter and cool in summer without any need of industrial heating or cooling, *etc.*

It is clear that this is a real revolution implying that many of today's business paradigms will disappear and new concepts have to be devised in order to generate new businesses. Every time more and more "voices of prophets" are claiming that humankind is running to destruction unless urgent measures are adopted in that sense. So, is this the only way to proceed?

6.3 Lean Design

Lean concepts were derived initially from the "Toyota Production System" (TPS) (Watanabe 2007; Liker 2006; Ohno 1988; Monden 1987), which in simple terms is defined as: "producing what is needed, when it is needed, in the time that is needed, with the minimum amount of resources and space." The whole objective of lean is the elimination of waste which is good enough to achieve an isolated success within a manufacturing company but it is not sufficient in terms of creating value in a sustainable way. What is needed then is a new paradigm that will take the *lean manufacturing* and *lean thinking* concepts from waste elimination into value creation. In order to make a significant change in enterprise performance and save ultimate system costs there is a need to have the entire enterprise undergo a lean transformation (Murman *et al.* 2002).

Waste in lean philosophy (also known as "*muda*" from the Japanese) means anything that does not add value to the user/customer; lean identifies seven categories of waste:

- *Transportation.* Product movements that are not actually required to perform the processing
- *Inventory.* All components, work-in-progress and finished product not being processed
- *Motion.* People or equipment moving or walking more than is required to perform the processing
- *Waiting time.* Waiting for the next production step
- *Overproduction.* Production ahead of demand
- *Inefficiency.* Low production rates due to poor tooling or weak processes
- *Reject/defects.* The effort involved in looking for and fixing defects

Besides "lean waste," real waste in environmental terms can be added following the discussion from the previous point.

From this scope, design pays a key role as it can be said that it has a very important impact on the way "lean manufacturing" principles can be applied and even on whether they could be applied at all. A product which is designed together with its manufacturing processes following the lean philosophy will be much more suitable for being "*leaned*" that any other product designed in the traditional paradigm. That is what is known as lean design.

Lean design is going to be an important part of this lean transformation, as up to 80% of the manufacturing cost is determined in the design stage. It is important to note that a product not "lean-designed" cannot easily be "leaned out" in the production stage. Hence the production of affordable and sustainable products would require an effective lean design and engineering.

Manufacturing companies really need a new model beyond lean manufacturing to ensure the transformation of the enterprise into lean environment. New trends in customers and market demanding value creation incorporating sustainability, cultural aspects, and customization have to be addressed by new concepts in business management.

Within this scope, a significant change in enterprise performance comes from the adoption of lean thinking throughout the entire product life cycle (Fig. 6.2 illustrates a view of product life cycle).

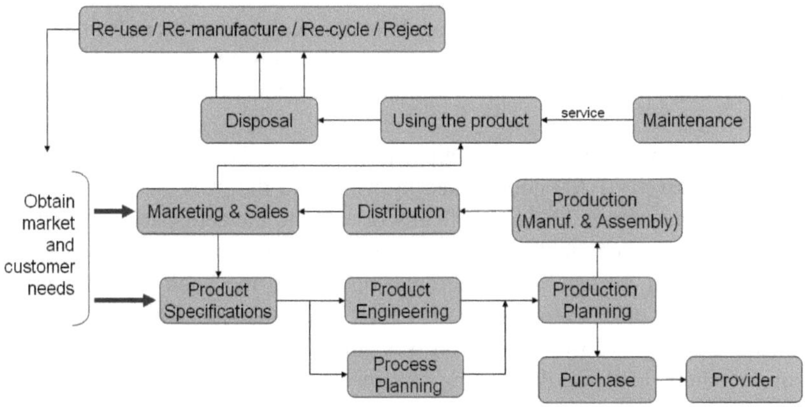

Fig. 6.2 Product life cycle overview

Waste in terms of "lean manufacturing" is considered as those activities that do not add value in the eyes of the customer. In applying lean concepts a major objective is to identify value and non-value added activities. This could also be adopted for product development where any activity that would result in customer requirements being met could be considered as value added activity. Product development activities must be formalized and structured in such a way that any engineering decision making is based on proven knowledge and experience. Failure to apply proven knowledge and experience will result in non-value added activities as product and process redesign (waste of valuable resources).

Background. Lean principles proposed by Womack (Womack *et al.* 1991) based on "Toyota Production System" (TPS) to improve the productivity of the shop floor by eliminating waste, may follow this sequence:

1. Specify value
2. Identify the value stream and eliminate waste
3. Make the value flow
4. Let the customer pull the (value) process
5. Pursue perfection

The lean principles have been applied to various manufacturing industries such as aerospace, consumer products, metal processing, and industrial products (Spear and Bowen 1999) and there is a significant amount of literature available, both commercial and academic, that detail the improvements organizations have made by applying lean concepts to their manufacturing facilities.

Previously mentioned concurrent engineering (CE) practices (Sect. 3.3.2) have been one of the major approaches to support product development; however, its applications are biased towards technology deployment and working practices as opposed to "value identification" and "value stream management" (Haque and James-Moore 2004). Karlsson and Ahlstrom (1996) carried out research based on observing several industries to come up with recommendations about the path to lean product development. The research did not define the meaning of lean and the general recommendations were more related to CE applications such as supplier involvement, cross-functional teams, simultaneous engineering, and integration of activities.

There have been two major lean thinking projects in USA and UK: the "Lean Aerospace Initiative"[7] (LAI) coordinated by MIT[8] and the "UK Lean Aerospace Initiative" (Harrison *et al.* 2002) specifically oriented to the aerospace industry. In the USA-LAI, launched in the 1990s, the efforts started by understanding the "Toyota Production System" (TPS) through publishing the book "The Machine that Changed the World" (Womack *et al.* 1991). The book gave a name to TPS as "lean manufacturing". Most of the effort was put into understanding lean applications on the shop floor and developing both practical models and the lean techniques to help the implementation. This effort then evolved to the lean transformation of the enterprise. This is now called the "lean enterprise" that covers the adoption of lean thinking to the management of the enterprise as well as its supply chain. Figure 6.3 illustrates the lean journey of the USA-LAI.

The UK Lean Aerospace Initiative was run from April 1998 to the end of 2001 to support member companies in meeting their improvement objectives and to establish an expertise and resources for the UK aerospace industry. The work emphasized lean manufacturing applications. Some part of the work addressed the issues of adopting "lean thinking" in new production introduction.

Several works related to lean design or lean product development have been achieved and published (Ward 2007; Huthwaite 2007; Mascitelli 2007; Morgan and Liker 2006; Fiore 2005; Kennedy 2003; Sobek *et al.* 1999). Some are based on research carried out in USA to observe and analyze the Toyota product development system.

These works propose some common pillars as a base for lean design:

- *System designer entrepreneurial leadership.* A technical leadership paradigm that efficiently brokers the right knowledge into the right product
- *Set-based concurrent engineering.* An exploration paradigm that generates extensive knowledge from many perspectives to maximize product alternatives with minimal risk

---

[7] http://lean.mit.edu/

[8] MIT: Massachusetts Institute of Technology. http://web.mit.edu/

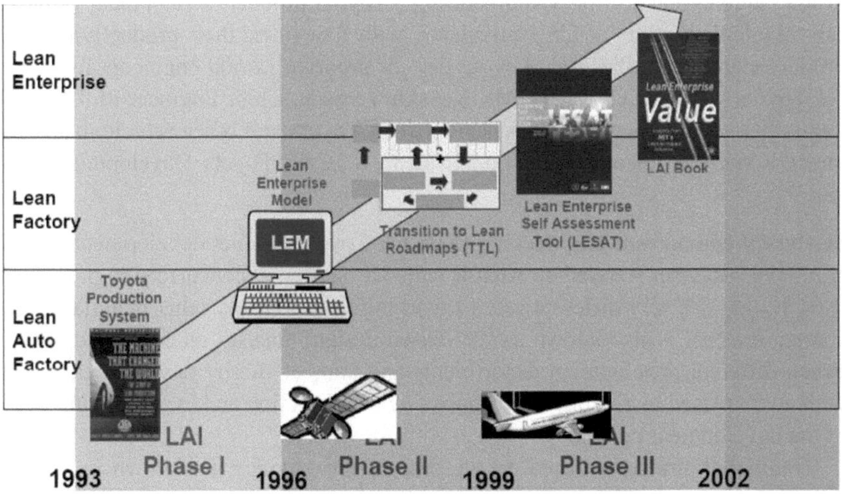

Fig. 6.3 Lean aerospace initiative (jointly developed by MIT and Warwick University)

- *Responsibility-based planning and control.* A management paradigm that provides efficiency, flexibility, and knowledge as the backbone for project execution
- *Expert engineering workforce.* A paradigm that assumes engineers have both the technical capability and access to the right knowledge to make the proper decisions to optimize the current product while building the knowledge for future products

Mascitelli (2007) based his book on his wide experience as consultant in product design in many companies. His approach is to provide a toolbox of methods that enable manufacturing cost reduction to become a foundational part of product design and development. Fiore (2005) attempted to merge lean manufacturing with six sigma to develop a template of three main foundation pillars:

1. The lean design
2. The manufacturing process
3. The control pillars.

Huthwaite (2007) put his experiences as consultant in design for manufacturing and assembly (DFMA), process control, and cycle cost into a new approach to provide designers with recommendations on how to avoid waste and to create values in their design.

In the USA, several researchers such as Durward Sobek II (Sobek *et al.* 1999; Sobek 1997), Ford (Ford and Sobek 2005) and James Morgan (Morgan and Liker 2006) made an effort to study the Toyota product development system. According to the National Centre for Manufacturing,[9] Toyota product development projects can take half the time of US equivalents, with four times their productivity (150 product engineers utilized by Toyota per car program *vs* 600 engineers for twice the Toyota's time at Chrysler). Mr. Kosaku Yamada, Chief Engineer of Toyota's Lexus line said: "The real difference between Toyota and other vehicle manufacturers is not the Toyota Production System; it is the Toyota Development System."

Set-based concurrent engineering (SBCE). Toyota product development (Ward *et al.* 1995) system is based on what is called a "set-based concurrent engineering" (SBCE) that is a very different way of working from so many other manufacturing companies. SBCE also known as "set-based design" focuses on collaboration between different development departments and aims at shorter development times with an increased quality level by improving collaboration and by paralleling parts of the development process.

Traditional strategies based on the design freeze-policy are known as "point-based concurrent engineering" (PBCE). PBCE advocates early selection of the supposedly best alternative in order to be approved ("frozen") by the top management. For sure the objective is to reduce development time and limit costs but, surprisingly enough, Toyota (as mentioned before) is capable of achieving the best-in-class results both in time and efficiency by using SBCE principles.

SBCE is set in practice by reasoning, developing, and communicating amongst design participants about sets of solutions running in parallel and relatively independently. As the design progresses, they gradually narrow their respective sets of solutions based on additional information from development, testing, the customer, and other sets of stakeholders as can be seen in Fig. 6.4. As they narrow, they commit to staying within the set(s), barring extreme circumstances, so that others can rely on their communication.

SBCE processes start with large design alternatives covering broad design spaces and then gradually narrowing the set of possibilities to converge to a possible design by eliminating the weakest alternatives rather than choosing one "best" alternative. It is a counter-intuitive approach and looks paradoxical to people trained in the traditional point based approaches. Various sets of alternatives are taken ahead for all parts of the product and the weakest ones are eliminated as the product development life cycle moves forward. SBCE assumes that reasoning and communicating about sets of ideas leads to more robust, optimized systems and greater overall efficiency than working with one idea at a time, even though the individual steps may look inefficient. This approach may require more time at the early stages to define the solutions but later stages can then move more quickly toward convergence and ultimately production, relative to more point-based processes (Bhushan 2007). Another advantage of this approach is related to the possi-

---

[9] http://lpdi.ncms.org/

bility of having more options for change face the uncertainties inherent to any new product design process (Ford and Sobek 2005).

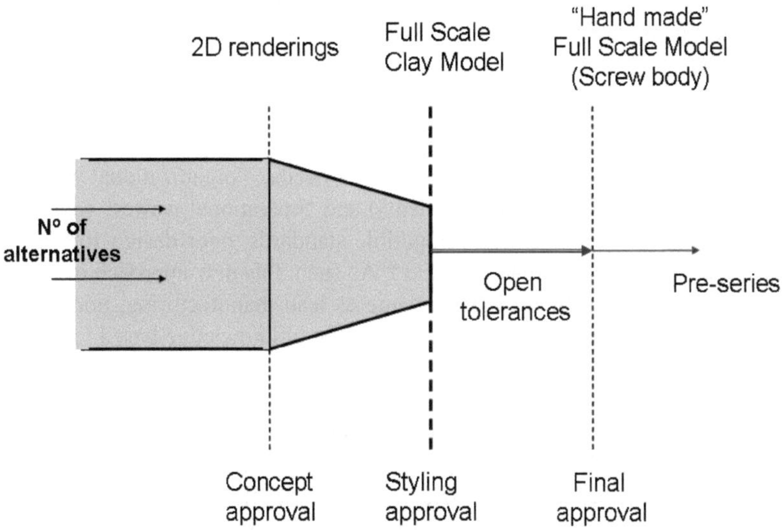

Fig. 6.4 Toyota's body development system (SBCE) (adapted from Ford and Sobek 2005)

Although SBCE has been known for many years and several research publications have described the process, it has not been picked up by many companies. The main reason may be that its principles are fairly counter-intuitive and, given the fact that in industrial organizations there is usually a time and budget constraint, the trend is going quickly for one design in order to demonstrate to the top management that the development project is on the right track. The information, decision, design, and organization complexity also increases as SBCE is a process requiring a very strict discipline for everyone to follow as there is no central control; it creates a self-organizing system. Further, the SBCE principles do not describe specific methods, techniques, tools, or frameworks for execution. SBCE existing literature does not provide information about the tools and techniques of the approach and does not address the consideration of lean thinking.

Another important characteristic of SBCE approach that actually hinders its wide use in industry is the need for high flexibility of the working force. In Toyota's case, launch deadline is immovable, so requiring early concentration of efforts enabling one to deal with as many alternatives as possible. This flexibility is fairly accepted and assumed by Toyota workers whereas it is not so easy to be implemented in other countries, other cultures.

New approach: lean design. The new approach proposes that the value creation and cost reduction activities become a foundational part of the entire product development system. Such a system would have two major types of wastes:

1. Waste associated with the processes of the product life cycle (Haque and James-Moore 2004)
2. Waste embodied in product design itself (*e.g.,* excessive complexity, poor manufacturing process compatibility, many unique and custom parts) (Mascitelli 2007)

Haque and James-Moore (2004) attempted to define different type of wastes in product development such as "strategy wastes" (*e.g.,* too many projects, inappropriate processing, poor understanding of customer needs); "organizational wastes" (*e.g.,* roles not clear, poor team arrangements) and "operational wastes" (*e.g.,* information formats, lack of common/compatible standards, poor design for "x" – manufacture, assembly, cost, reliability, *etc.*). As such, this new approach of "lean product (and process) design" is not the same as lean manufacturing, nor its replacement. It is a new concept to enable the creation of knowledge-based factories going beyond lean manufacturing, considering the entire product life cycle in a manufacturing enterprise and its implementation in real scenarios.

New models for product/process development have to be developed. These models will be based on lean thinking and will consider entire product life cycles, providing a knowledge-based user-centric design and development environment to support value creation to the customers in term of innovation and customization, quality as well as sustainable and affordable products.

Performance measurement considering human resources, technology factors and processes of an enterprise has to be used to measure the readiness and level of adoption of lean thinking principles. This will lead to an understanding of how product and process development is structured and what is needed to streamline the process to maximize value creation. Hence, the mapping of product and process development is addressed to measure the values from the customers' point of view and estimate life cycle costs, including the manufacturing and in service components.

The new approach will enable manufacturing companies to balance the need to react to value creation opportunities and the efficiency to deliver them effectively. This will be achieved as any engineering decision making will be based on proven knowledge and experience, to reduce risk and maximize utilization of resources of both the enterprise and its supply chain.

Eco-innovative lean design. Combination of eco-innovative principles as discussed in Sect. 6.2 and lean design can constitute a sound paradigm for designing products in the twenty-first century. Eco-innovative lean design can cope with the requirements arising from today's increasing demand for customized products with short delivery times. Furthermore, products and services should be capable of fulfilling users' demands, while also reducing total life cycle costs and environmental impacts.

6.4 Open Innovation

There is a clear shift from closed to open innovation, which is likely to continue. The specific challenges are imposed upon industry leaders that play the role of an "industry architect." In "Open innovation in systemic innovation contexts" (Maula *et al.* 2006) it is indicated:

> While the creation of systemic innovations is frequently steered and to some extent controlled by these industry leading firms, questions that rise are what role other firms in the systemic innovation process can play and what the unique challenges are for these players. Future research may investigate how producers of complementary products influence the evolution of systemic innovations for instance to position their complementary innovation as centrally as possible in the system.
> An additional interesting question would be to which extent small and medium sized enterprises (SME) can play the role of architect steering the evolution of a systemic innovation. Many of the mechanisms discussed in this chapter might be unavailable for SME. At the same time alternative open innovation processes such as open source development (Grand et al. 2004; von Hippel and von Krogh 2003; Kogut and Metiu 2001) might open alternative avenues for SME to steer open innovation processes but these avenues would need to be investigated further.
>
> Maula *et al.*

While a framework for resource allocation in a systemic innovation context has already been investigated, the mechanisms of resource allocation available to firms in an *open innovation* model are still to be investigated (*e.g.,* internal decision making within open innovation). It still needs to be investigated how firms reconcile the some times conflicting demands, time horizons, and resource allocation mechanisms between internal and external resource allocation and thereby further develop the Bower-Burgelman model (Burgelman 1983a, b; Bower 1970) to incorporate external resource allocation (Maula *et al.* 2006). A possible approach for such SME driven innovation is discussed in Sect. 6.5.

6.5 Innovation in Non-hierarchical Networks

As discussed in the previous chapters, collaborative networks nowadays are appearing in a number of manifestations, including virtual organizations and enterprises, dynamic supply chains, professional virtual communities, *etc.* which has led to intensive activities on their modeling. Specifically, company networking is seen as a powerful "tool" for innovation in the twenty-first century. The observed aspects spread over a number of areas, such as computer science and engineering, communications and networking, management, social sciences, law and ethics, *etc.*, contributing to define and characterize emerging collaborative organizational forms. Collaborative networks are likely to continue to be a key approach for modern innovation.

6.5.1 *Virtual Breeding Environment*

An effective creation of dynamic collaborative networks requires a proper context in which potential members are prepared to rapidly get engaged in collaborative processes. The concept of "Virtual Breeding Environment" (VBE) (Afsarmanesh 2005) has emerged as an important facilitator for wider dissemination of collaborative networks and their practical materialization. The concept of "Virtual Breeding Environment" is defined as "cradle for dynamic and agile establishment of opportunity-driven collaborative networks" and "regulated open, but controlled-border association of its members."[10]

VBE form is a promising new model to support innovation in modern networked industry. It may be seen as a complementary combination of:

- *Business networks*
- *Knowledge community concepts*

Such a combination has already been considered in Sect. 5.4.2, where a relatively simple ICT solution to support such SME networking is briefly described, specifically related to the construction industry. However, VBE is an advanced concept for company networking based on such combination. VBE are ecosystems which require effective solutions for interaction with their business, economic, legal and ecologic environments and effective support for network/community management as an appropriate answer to the high dynamics of the business conditions. Therefore, ICT is a key facilitator of such VBE models. It can be stated that the complex multiple relationships of modern networked industry with business, economic, legal and ecologic environments can be optimally managed within a VBE concept, which opens new opportunities for business models unthinkable without advanced ICT systems (Expert Group 2005). VBE are likely to be specifically appropriate to support innovation processes in networked SME in accordance with the above described open innovation principles.

VBE can be realized by the application of different digital enterprise technologies, which enable new dynamic networking in both vertical integration among business processes and horizontal integration for internal (within network) resources planning and process scheduling. Virtual breeding environment allows establishment of innovative interactions of such horizontal or vertical networks with a knowledge community. As indicated in Chap. 5, modern industry, and especially small and medium sized enterprises (SME), require highly dynamic and flexible cooperation models which effectively combine business networking and knowledge community (e-Mult 2006). VBE is likely to be an appropriate response to the need to establish closer business collaboration as a conjoint alliance of *industry (SME) and research and technological development entities (RTD)* enabling creation of integrated teams that will successfully and profitably cope with challenging

---

[10] ECOLEAD project: European Collaborative Networked Organizations Leadership Initiative. http://ecolead.vtt.fi/.

complex innovation targets. New business networking will assure higher efficiency of the co-operation and integration processes and development of new products and services with a high added value for clients.

Therefore, key issue of VBE approach is an effective combination of *business networking* and *knowledge communities* enabled by advanced ICT platforms.

Appropriate business models for such a VBE approach can be based on the basic concepts of network as a form of relationships among actors and activities/resources (Seppola 2004). Transformation of this approach to VBE as "a self-organizing digital infrastructure aimed at creating a digital environment for networked organizations that supports the cooperation, the knowledge sharing, the development of open and adaptive technologies and evolutionary business models"[11] is presented in Fig. 6.5: the key problems of "exchange relation" among actors in a network and "control of activities and resources" are to be solved by an appropriate ICT platform for effective communication and knowledge sharing among actors and by innovative software services.

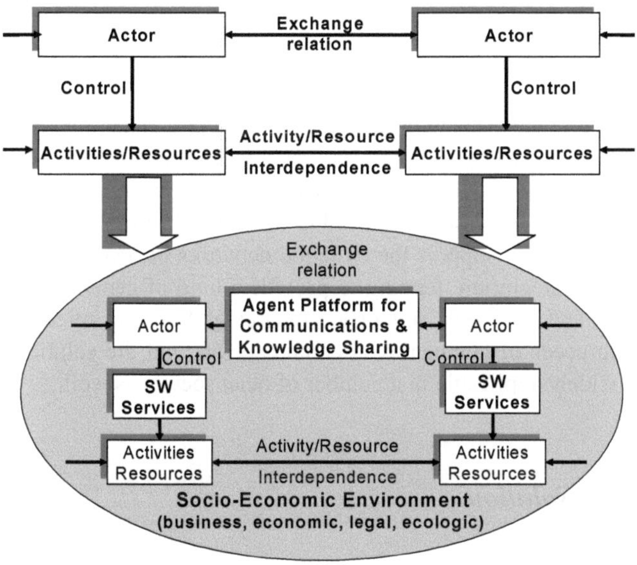

Fig. 6.5 Concept for management of network (Urosevic and Stokic 2007)

In a variety of business network classifications dealing with Virtual Business Networks (VBN), specifically important for innovation in SME-driven networks is the one addressed in VE-Forum.[12]

Among the three groups of networks identified according to the network topology, the *peer-to-peer network* topology is likely to be the most appropriate for

[11] www.digital-ecosystems.org

[12] http://www.ve-forum.org

modern companies. This model entails mutual relationships between all (or most) network partners without a prominent strategic centre (polycentric structure). Instead, initiatives can be launched from each node of a network. Generally speaking, such typical *non-hierarchical networks* seems to be appropriate in industries where access to knowledge and expertise is of primary concern and where a high dynamics is to be expected, which in turn requires high organization flexibility and scalability (collaboration form, introduction of new partners, content adaptability, *etc.*) (DeVries and Kommers 2004). Such a model is appropriate for SME intending to organize their collaborative innovation process which is not "dictated" by a large company ("industry architect" as indicated in Sect. 6.4). Therefore, a VBE approach combined with concepts from the approach of "Competence Cells Networking" (Ackermann and Mueller 2007), as a component of the "non-hierarchical production networks," appears to be the most appropriate form for virtual collaboration of SME, and the peer-to-peer topology is a promising model to be applied as a basis for VBE for SME driven innovation. This model allows for effective combination of business networking and knowledge community, needed for open innovation driven by SME.

The VBE, as indicated in Fig. 6.5, is to be enabled by the system composed of ICT platform and SW services. Scalable, open architecture platforms providing the possibility to interface the system with different dynamically changing environments (both within SME networks and with "external" environments) and aiming to achieve seamless data exchange among networked SME are needed. SME are enabled to integrate their processes and by this to establish more effective partnerships. Such ICT platforms support the dynamic networks of SME in both the vertical and horizontal integration, tending to provide a kind of generic VBE system component. VBE and ICT tools supporting (enabling) VBE are often focusing upon the specific needs of companies in certain sectors, but the solutions are often tending to be widely applicable in a number of other sectors as well.

6.5.2 Agent Based Solution

ICT platforms to facilitate virtual breeding environment could be implemented using different technologies. One such implementation is using software agent technology and is investigated in the scope of the project e-Mult (2006). This represents a more advanced ICT solution for a combination of business networking and knowledge communities with respect to the one presented in Sect. 5.4.2. It allows for an efficient implementation of the VBE concept as a very promising approach for innovation in modern companies.

Agent technology appears to fit the requirements of a highly dynamic system as this one is. Strohmaier *et al.* (2007) take an agent-oriented modeling approach, proposing agents as knowledge transfer instruments. Also, Salles and Furtado

(2004) describe an initiative which proposes the construction of a general knowledge system based on agent communities. [13]

The functionalities to be realized by agents are presented in Table 6.2.

Table 6.2 Agents for virtual breeding environment (e-Mult 2006)

Agent type	Application within virtual breeding environment
(Market driven) negotiation agents	Provide new abilities in terms of reaction to changing market and other conditions, and to have the flexibility to raise and lower trade expectations and requirements from different networks partners
Monitoring agents	Watch the external and legacy systems for different information (*e.g.*, market related, legislative aspects, stocks, resources) and to monitor parts of the VBE system. A user is enabled to create monitoring tasks[14]
User interface (UI) agents	Assist user of VBE services, providing the properties of adaptability to user's skills, interface sharing with user and support through the learning effect and adaptation to increase user's performance
Input validation agents	Check user inputs received from the UI agents (*e.g.*, entered knowledge, problems, *etc.*), request the UI agents for missing user inputs, *etc.*
Management agents	Handle the communication between user interface and wrapper/business logic-agents. Responsible for the distribution of work, collection of results
Wrapper agents	Allow an agent to connect to a (non-agent) ICT system uniquely identified by software description. The business logic of the VBE system, the data in existing ERP-systems and other systems can be accessed[15]
Ontology merger agents	Deal with the tasks emerging from using a common ontology, which guarantees an efficient communication of the systems and actors in VBE, *i.e.*, a common ontology maintenance (see Sect. 4.3)[16]

The multi-agent platform allows establishment of a set of software services needed (Urosevic and Stokic 2007; Große Hovest *et al.* 2008), *e.g.*:

[13] The agent platform for VBE should be in compliance with the IEEE-FIPA standards for software agents (including all necessary components for an agent platform), what is vital for ensuring interoperability of autonomous agents. The software development framework JADE (Java Agent Development Framework) can be applied aimed at developing multi-agent systems and applications conforming to the standards, which is also the runtime environment that provides the basic services.

[14] Monitoring agents may: (1) process watching/monitoring tasks created by a user; (2) present gathered information to the system's user; (3) handle the notification level, *e.g.*: present information when the user requests it and notify the user automatically about information.

[15] The advantage is to enable existing IT systems to "behave" like agents, while hiding the real implementation.

[16] These agents are to analyze frequently the repositories of the participants in order to detect and indicate definitions worth being added to the global ontology. On affirmation it will then adapt the new definition at the common ontology and introduce this change to the repositories where needed. Agents are maintaining and improving the interoperability of the collaboration network.

- To set-up SME *business or knowledge network(s), i.e.,* services for dynamic networks set-up
- To *operate innovation within business networks, i.e.,* services for dynamic networks management
- To support knowledge sharing infrastructure within the *knowledge communities* (*i.e., knowledge forum)*

Such a platform is oriented to enable primarily the peer-to-peer network topology, addressing many different aspects relevant for industry such as protection of confidential company information, mobile workers, *etc.*

Various types of VBE services can be envisaged in the solution described: agents to enable effective management of interactions with business, economic, legal, and ecologic environments in which a VBE is operating, and agent-supported services to set up and operate networks and to manage a *knowledge community*, as illustrated in Fig. 6.6.

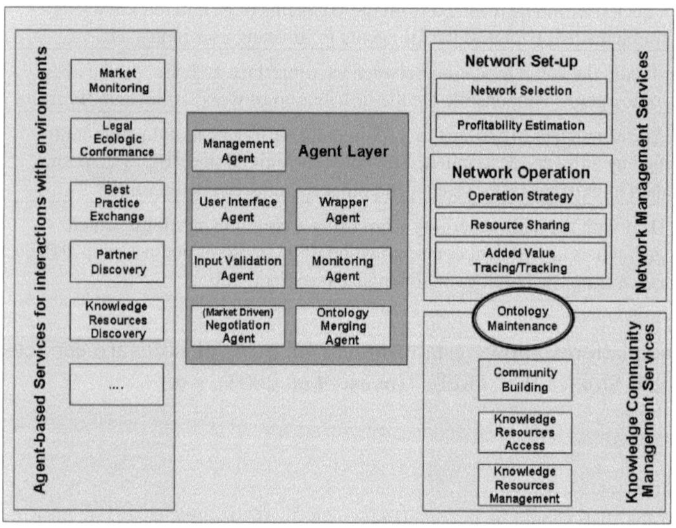

Fig. 6.6 Agent based platform for virtual breeding environment (e-Mult 2006; Urosevic and Stokic 2007)

Services for set-up of dynamic networks comprise services for optimal network selection, based on a set of enterprise models with structured definitions of dynamic networks and services for basic profitability estimation, using the information from the agents.

Services for management of dynamic networks comprise, e.g., services for operation/innovation strategy support and innovation scheduling, services for decision making support on resources, and services for measuring and tracking of added value within the network.

In addition, virtual breeding environment may include other functionalities to manage such networks (logistics, stock handling, *etc.*), but the above three are identified as the most critical for an effective operation of virtual breeding environment aimed at innovation led by SME in industry. The second important part of the proposed virtual breeding environment is the knowledge forum described below.

Services for decision making on innovation strategy and scheduling of the activities have to support management of innovation processes within the network. The services have to cope with the critical problems within virtual breeding environment in industrial domain: uncertainty of market, multiple, dynamically changing networks requirements, *etc.*

Services for dynamic sharing of resources and distribution of activities within the multi-threaded networks relate to the global management of networks through the resources sharing and distribution of work within complex networks of SME.

The services address, *e.g.,* the key scenario of global distribution of work/resources among SME to provide sufficient amount of resources needed to allow for innovation. The functionalities for innovation activities planning (optimal distribution of work among companies at a global level), and for optimal joint usage of resources (equipment and labor resources), have to be provided.

The objective is to provide functionality to cover different time horizons depending on specific network needs. The main problems are the multi-threaded character of networks to be taken into account by the algorithms and difficulty in the prediction of market demands which may change very abruptly.

Services for measuring and tracking of added value along chains support solution of the critical problem within networks: how to measure added value in such networks along the chains and product life cycle. The problem is how, based on measurements of effort spent and investments in different processes and all other costs on one side, and based on the price achieved at the market on the other, to determine the added value which these processes can reach. The objective is to have a service to observe dynamically (on-line) added value within the (parts of the) network. The benefits of such a tool are obvious: it enables SME to identify the real added value of different products and processes and thereby to define a better business strategy and better manage their networks, both in the short term (fast reaction on changes in conditions) and the long term.

Knowledge Forum (KF), *for the knowledge community,* is a place where SME are in a position to identify effectively potential innovative applications of their products and processes needed for such applications, identify how to set-up value chains (using developed services), discover the most suitable partners, establish common and "members only" areas for knowledge sharing within networks, and "meet" RTD partners and their ideas, *etc.* Furthermore, the system provides relevant knowledge related to legislative and standardization issues, regularly updated by the agent-based service legal/ecologic conformance (see Fig. 6.6).

KF includes a set of tools/services for knowledge management (KM), representing so-called KM infrastructure (Thie and Stokic 2001). Based on the analytical forum model (Hendryx 2003), the concept of a general functional architecture of KF provides the following services (Fig. 6.7):

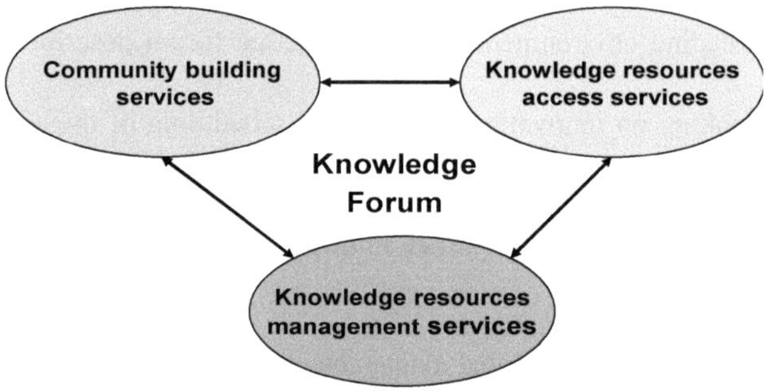

Fig. 6.7 General functional architecture of the KF (e-Mult 2006)

- *Community building services.* They support the processes of knowledge community building by providing the instruments to foster professional interaction and socialization. Different tools (*e.g.,* discussion forums and weblogs) are tailored to the end-user needs and strongly integrated with the KM infrastructure. Other general communication and information dissemination tools complement these services.
- *Knowledge resources access.* These services comprise the fundamental set of functionalities in KF, allowing for creation, searching, and update of knowledge resources. Although much of the community information/knowledge will be created in communication/interaction processes, there will also be the need to create/access knowledge in a more structured way. Digital content and document management are the natural approaches related to this issue.
- *Knowledge resources management.* It is a set of infrastructural functionalities that support information and knowledge acquisition, organization, and storage.

6.6 Trends in Collaborative Innovation and Collaborative Working Environments Technology

Collaborative innovation is obviously a key aspect of the innovation process in modern industry as explained in previous chapters. ICT supporting collaborative work are and will remain the key means to facilitate such collaborative innovation processes in industry. There is a consensus that it is time to move the focus towards the improvement of the way in which people innovate together, *i.e.,* to research and develop new working environments ready for collaborative innovation. They should be developed in line with the new ICT trends, the so-called third wave of Internet, *i.e.,* utility-like network, sensors, and wireless technologies, and commodity-like software (Laso-Ballesteros and Salmelin 2005; NWE-EC 2005).

In the next few years, collaboration technologies will require models of the dynamics of human interactions that can simulate behaviors, characteristics, and appearances to simulate physical presence. Behavioral and social experts will be essential members of development teams for such collaborative working environments (CWE) solutions.

Future collaborative technologies have to ensure integrity, persistence, security, and privacy beyond physical data, because co-workers will be dynamically connected in a global world through heterogeneous means and will use various mobile devices at several places. From the application-centric and process-oriented point of view, collaborative ICT have to be customizable to different communities. They have to be flexible for the users and, under corresponding circumstances, they have to be usable for professional and private environments, especially supporting the collaborative open innovation approach as described in Sect. 6.4. They have to be able to identify the most relevant tasks for a worker, taking into account her/his context and her/his needs for mobility and for collaboration with other workers. They should be loosely coupled with management and stream work of inter organizational processes and processes across disciplines.

For these reasons, next generation "upper layer" Collaboration@Work middleware for distributed environments:

- Mediates between the different layers
- Leaves the diversity visible for the user
- Hides (in contrast) the implementation details.

The vision of CWE (Expert Group 2006) assumes that the new collaboration spaces should not be any more strictly application-oriented. They should directly support humans in their activities – without asking humans to think on specific ICT applications such as e-mails, *etc.* (Prinz 2005) – but providing resources in activity context and direct interaction. Infrastructure should support users in organizing, structuring and securing the work on innovation based on a diversity of best practice collaboration patterns (control of performance, right storage, patterns, *etc.*).

In line with the open innovation approach described above, the new collaboration spaces should not be restricted on workplaces; they should apply to *ALL collaborative spaces*. Such collaboration spaces must be *pro-active, goal oriented* environments to help individuals, groups, communities, and organizations to solve the problems in an intelligent way, reach goals, *etc.* ICT tools will have to support social activities and relations and they must be culture aware environments (human cultures, group cultures, corporate cultures, languages).

The future CWE for innovation will address (Expert Group 2006):

- Ontologies for collaboration at work
- Plug and play interoperable service oriented architecture (SOA) for collaboration at work
- Smooth "upper layer" middleware interaction with the underlying layers
- Interaction among peers (workers, systems, robots)
- Utility-like computing capacity and connectivity
- Contextualization and content
- Group-level security, privacy and trust
- Mobility at work
- Reference architecture for collaboration at work (as explained in Sect. 4.2.4)

In the near future, new areas for e-Collaboration technologies will involve exploring more sophisticated application domains with a view to boost innovation in the business ecosystem. Among these future lines, it is worth mentioning:

- Collaboration technologies for knowledge activation
- Collaboration technologies for applied collective creativity.

Such solutions will make use of networked devices embedded in any terminal and product, which will allow continuous, seamless streaming of communications, content and services-exchanged among workers, artifacts, and their partners and customers.

6.7 Semantics for Collaborative Innovation

One of the most promising approaches to support collaborative innovation is to use "semantics" inherent in collaborative work. The collaboration within innovation processes in modern, highly flexible industry, operating in global economy, is *"dynamically changing,"* since the collaboration patterns, teams, organizations, *etc.*, have to be changed often to face new problems and conditions. Dynamics of collaborative work in industry appear at two levels:

- *Real-time dynamics.* During a specific collaboration process the situation is often dynamically changing (*e.g.,* changes within teams, location change/mobile work, *etc.*)
- *Higher-level dynamics.* Due to required high flexibility of modern industry, "forms" of collaboration are often changed (new partners, rules, relations among partners, *etc.*)

Knowledge management (KM) systems, as a key means for effective collaborative innovation, have to support such dynamics. The assumption is that if, *e.g.,* a KM system supporting a group to solve certain problem or developing an innovation would have knowledge on previous "similar" collaborative work (in "similar" context), it would help in sharing knowledge and solving the current problem. As explained in Sect. 5.3, "similarity" has to be defined for a specific company, Extended Enterprise, or company network. It may include "similarity" on technical topics which a specific problem addresses, but also on collaborative context:

- Number of people involved
- Locations of collaborative work
- Collaboration patterns, *etc.*

One of the most critical problems within KM for collaborative work in industry is how to acquire/provide knowledge, efficiently and promptly (on-line), on real-time dynamically changing collaborative work needed for optimization of KM, and how to use such information effectively to support dynamic collaborative work. Basic assumption of virtual collaboration is the availability of digitalized information, which is not the case for social interaction in the physical world. Advanced ICT environments provide a number of services to support collaborative working in the virtual world, as described in Chap. 5. However, the observation of social interactions in order to provide knowledge in an appropriate form is a complex task due to numerous aspects such as it always needs to be mediated (*e.g.,* by multimedia systems), time aspects, IPR and privacy issues, *etc.* On the other hand, collaborative work on innovation in industry often occurs as a combination of *virtual and physical interaction.* Therefore, monitoring and recording of collaborative work has to be effectively solved *without requiring teams* to document explicitly their actions, taking into account both *physical and virtual collaboration* (and their mutual relations and smooth transition from one pattern to another).

Once the collaborative work is monitored and documented, the problem is how to extract the knowledge which could be useful for future collaborative work and how to reuse this knowledge for collaborative solving of the current problem, taking into account *dynamical* changes in collaboration (on both above-mentioned levels). The objective is to *exploit semantics embedded* in structured and unstructured content produced/used by participants during collaborative work, in order to provide more effective KM for collaborative work. While structured content can be readily processed (*e.g.,* data strictly formatted according to the interface of a collaborative service), unstructured content (information provided by ambience sensors, text transcripts of conversations, meetings video files, drawings made during collaboration sessions, *etc.*) is harder to process, but might sometimes hold more useful information than the structured data (hidden intelligence). However, even records automatically collected in a structured way (*e.g.,* by traceability services) may have very different granularity and amount of detail compared to what the KM system actually needs. On top of that, a high dynamics in collaborative work (at both above-mentioned levels) impose many challenges regarding context modeling, extraction and re-use.

6.7.1 *Key Technology for Semantics for Collaborative Innovation*

Reusing context information for improving collaborative work. Using context information (for context-aware or ubiquitous computing) is an active area of research for future innovation processes, with various context capture methods and context languages defined.[17] The current research on knowledge context is oriented toward active knowledge delivery based on context or capturing and utilization of contextual knowledge, but such initiatives have specific goals, so an intense study of collaborative work and its patterns is necessary to devise a suitable context model (Ning *et al.* 2007).

A knowledge-based context model is crucial for context-aware collaborative services. Since an ontology allows knowledge sharing, logic inference, and knowledge reuse, it is a widely accepted approach for context knowledge modeling. Several semantic specification languages such as Resource Description Framework (RDF) provide potential solutions for context modeling. Based on context ontology, logic based context reasoning can be realized such as consistency validating, "subsumption" (process of encompassing as a subordinate or component element) checking, *etc.* More important domain specific rules can be defined to infer implicit context from explicit context, and high level context from low level context. Other statistic and machine learning approaches can also be adopted for non-logic context reasoning.

Ambience Intelligence (AmI) to support collaborative innovation. AmI technology, promising to be a powerful approach to effectively solve "people issue," is still not introduced in industrial practice, but it is expected that in the near future it will be massively introduced in industry and specifically in the shop-floor environment, serving different applications, in order to support modern human-centered industrial concepts. AmI technology will involve new use of sensors (*e.g.,* wireless intelligent sensor networks) to observe both human and process behavior aspects, such as human interaction with machines/processes, material handling by human operators in highly flexible manufacturing, *e.g.,* by smart tags, *etc.* (Stokic *et al.* 2006).

The AmI applications in industry, although delayed as initial visions, are offering ever more advantages for new working environments and are initiating more and more new technology developments. As indicated above, the full application of AmI in industry is still to be achieved within the next few years. As explained in "Collaboration@Work: The 2004 report on new working environments and practices" (NWE-EC 2004), from the industrial perspective, a less human – and more system – centered definition of AmI is considered. However, as already in-

[17] Starting with the pioneering work at XEROX PARC, other notable frameworks are Context Toolkit (Berkeley), CAMELEON project, C-OWL and the Kimura System, TOSCA & Groupdesk, Virtual Office, Watson (Budzik and Hammond 1999).

dicated, the modern manufacturing concepts turn to *human-centered* approaches. Therefore, the application of human centered AmI technologies is promising to meet needs of such concepts effectively. The prices of components comprising AmI solutions are reducing and it can be expected that in the next decade a number of industrial companies will introduce different AmI technologies in industrial processes serving different purposes. Many issues still have to be solved in order to bring the AmI technology to industrial sectors, such as robust, reliable sensors, intelligent user interfaces, safety, security, *etc.*

Extraction of meaning from social interaction patterns. Automatic analysis of social interaction implies extracting knowledge from a data set which describes the social situation. Depending on the nature of the considered data, this process may be based on information provided by emerging (AmI related) technologies such as computer vision, speech recognition, Global Positioning System (GPS), or motion sensors. Various methods, such as pattern matching, labeling/tagging, are used at the lower level, but the limitations of extracting knowledge/meaning from multimedia content, related to social interaction patterns, derive from limitations of multimedia mining and the restricted ability of current techniques to overcome semantic gap.

Tagging of structured/unstructured information for collaborative work. Text mining has received a lot of attention in the last few decades and multiple techniques exist for classifying (tagging) this kind of information. Research is currently focused on tagging of multimedia information. Techniques such as computer vision and pattern matching try to bridge the "semantic gap" between low level features (pixels, lines, shapes) and high level ones that humans use to tag images (a beach, a sunny landscape). Various systems have been implemented. Proposed systems use tagging of both unstructured and structured content to enhance virtual collaborative work, but many fields (such as industrial innovation), with high dynamics of physical and virtual collaborative work, still need to benefit more significantly from such an undertaking. Automatic annotation techniques usually rely on ontology structure analysis and extraction patterns (Cosulschi *et al.* 2006; Castro Reis *et al.* 2004).

Semantic correlation and collaborative work. Semantic correlation has become very important and is usually associated with concepts such as semantic Web, semantic interoperability, and context-awareness. A lot of these issues are dealing with the problem of the same data having different semantic loads in different contexts. Latent semantic analysis (LSA) and probabilistic LSA have been the steps toward taking into account the semantic aspects of data.

6.7.2 AmI Based Solution

A possible solution for usage of semantics in dynamic collaborative innovation is a combination of AmI technology and services for management of social interactions (MSI). The future Collaborative Working Environments (CWE) will use ambient sensors and devices and services for MSI to "support" KM services for collaborative knowledge generation and sharing within dynamic, physical and virtual collaborative work. For example, within the project K-NET (2008) a solution focusing upon application of services for MSI to support KM services within virtual collaborative work is under development. A solution combining AmI technology and services for MSI still has to be explored and it has to provide answers to the questions on how to:

- Efficiently *monitor dynamic, combined physical and virtual, collaborative work* (without requiring teams to document explicitly their actions), using management of social interactions services and sensors (AmI technology) surrounding users, so that this knowledge can be collected, semantically enriched and reused for future collaborative work
- *Extract* context from the structured and unstructured content produced/used during collaborative work, provided by AmI technology and/or obtained by services for MSI (including those for monitoring virtual collaborative activities)
- *Reuse* such extracted context to support KM in new collaborative activities

By answering these questions, CWE will provide new services for management of social interactions, ambience sensing and for KM within *dynamic collaborative work on innovation.* These services will extract context from the content produced by humans, provided by AmI technology and by MSI services during the (past) collaborative work and use the extracted context to support KM within (future) collaborations.

The AmI systems will provide information (so called AmI information) that can be used to monitor collaborative work. One of the basic assumptions of the "ambient intelligence" is the principle of ubiquity, as a huge source of additional data/information. *AmI technology* can support collaboration among teams by providing information difficult to capture in digital form. Therefore, AmI systems will provide cost and time effective ways to collect a radically higher amount of information/knowledge and, quite important, to gather information/knowledge on *combined physical and virtual collaboration* which up to now was difficult to acquire, *e.g.,* knowledge on human operators, their environment and context of work, including business processes, information on individual preferences, behavior, experience of operators, *etc.* This will open new and up-to-now non-exploited opportunities to optimize collaborative work: information collected using AmI technologies will give a new insight into the performance of different actors and on their mutual interaction. Therefore, it may be stated that AmI for new workspaces in industry is essential to provide human centric collaboration.

The assumption is that a *synergy* of management of social interactions services and AmI solutions, serving different applications, may be used to provide information needed to enhance KM for collaborative work. The information collected using AmI and management of social interactions services (MSI) give a new insight into both physical and virtual collaborative work (and their interconnections) in industry and on interaction between different subsystems and their behavior under changing circumstances, allowing enhancing KM supporting such collaborative work.

As mentioned above, the collaborative environment currently discards a lot of information with potential useful semantic load. Therefore, future CWE will apply the context extraction which will follow the "collaboration patterns" approach, *i.e.,* the collaboration pattern models which are relevant for KM in collaborative work (*e.g.,* physical/virtual collaboration, asynchronous/synchronous work, location, time and iterations needed in solving problems, *etc.*) will be used as context models. Such models/ontology serve to extract context to provide actionable meaning from the collaborative activities. The MSI services and ambience sensing services will be needed to provide structured content according to the identified models. The unstructured content produced/used during collaboration and/or acquired by surrounding AmI systems have to be semantically enriched.

The services will, essentially, be built as a feedback loop (see Fig. 6.8): content gathered/used or produced during collaborative innovation activities, such as searches, meetings/chats, *etc.*, and provided by management of social interactions and ambience sensing services, will be processed and semantically enriched in order to improve searches, problem solving and decision making during future collaboration situations, *i.e.,* the extracted context will be used to improve "operation" of KM services. As the collaboration is dynamically changing in real-time, the context extraction and KM services improvement have to be carried out *dynamically in real-time.* On the other hand, as the collaboration is dynamically changing at a higher level as well, it is necessary to update models and ontologies (external loop) dynamically.

Therefore the conceptual structure of the proposed solution as illustrated in Fig. 6.8 is built around two loops:

- *An internal loop.* It contains the service infrastructure (*i.e.,* monitoring of collaborative innovation activities *via* management of social interactions and ambience sensing services) and real-time extraction of context and semantic upgrade of the content in order to support knowledge reuse. Whenever a new collaboration situation emerges, and the innovation activities are being monitored, actual context can be extracted from it. Based on this, it is possible to search previous situations, stored in the knowledge repository, that deal with context that is "similar" to the actual one and/or to search for semantically upgraded knowledge resources generated/used in previous situations. The service infrastructure includes:

- Ambience sensing services
- Management of social interactions services
- Collaborative knowledge repository (distributed knowledge basis in EE)
- KM services

Real-time extraction of context and semantic upgrade of content includes context provider and context model repository. Of course this internal loop includes different application specific tools which may be collaboratively used within specific innovation processes (*e.g.*, CAD[18] tool).

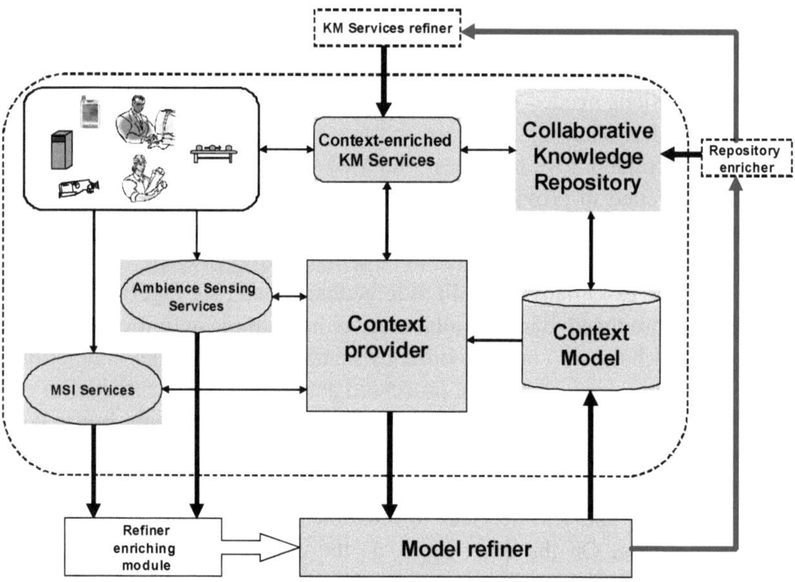

Fig. 6.8 Conceptual structure[19]

- *An external loop.* It performs the functionality of "dynamically" refining the context models and the KM services (due to the above-mentioned *high-level dynamics* in collaborative work). The key component here is the model refiner which will dynamically update context models, collaborative knowledge repository and KM services.

[18] Computer aided design

[19] Please note that, for the sake of simplicity, only the key interconnections/information flows are indicated in Fig. 6.8. Obviously, *e.g.*, context provider will use information from collaborative knowledge repository and KM services, what will make semantic upgrade of content generated/used, *etc.*

When users are working in collaborative environment, the context provider will continuously perceive their working status and provide the activity context, to the KM services. The KM services will use this to recommend knowledge to the users to support their current collaborative work.

Context model. The context modeling (CM) within this concept represents an abstract description of the monitored collaborative work (both physical and virtual), relevant for KM activities within these collaborations. It is based on an ontology whose concepts and relations are directly derived from the collaborative situation domain; the concepts and relations can also be seen as "semantic labels" for the instances stored in the collaborative knowledge repository. CM, the ontology and the knowledge resource are separated from each other, but they do rely on a triangle relation to one another, as each aspect from one instance is used in the other instances to fill data, tag it, expand, or chain it.

The model building can be based on collaboration patterns models, which are relevant for KM in collaborative work applying approach of information pyramid of virtual collaboration with different levels of information granularity (Biuk-Aghai 2004). These models will be then translated as meta-data into upgrades of the ontology to support KM services.[20]

Context provider. The context provider services play the central role in the system's inner loop. Actionable meaning has to be extracted from "raw" information from multiple sources across collaborative environment. By actionable meaning is meant machine interpretable knowledge that can be consumed directly by KM services to adapt themselves to support collaborative work better on innovation. Approaches for context sensing have to be applied: the context is seen from a dynamic point of view, as a constantly (real-time) changing set of context elements. The context provider module will then work with thresholds which show the granularity needed for the context-changing elements, *i.e.*, what is the atomic context-changing event, or the minimal event which is considered to change the context.

By using the context model and structured/unstructured information provided by users and management of social interactions, and ambience sensing services, context provider is able to process this information, to automatically annotate it and to store it in the collaborative knowledge repository, associated with actual contexts. Context provider is, therefore, used to add semantics (tag) to the content provided/used within the collaboration process (both content produced/used by the humans and content provided by the management of social interactions services and/or AmI). Learning capabilities of context provider are supported by the unique representation of concepts and semantic structures, which is facilitated by the un-

[20] Context model can be based on a new OWL-Based Activity-Centric Collaboration Ontology to represent the extracted information as an explicit machine interpretable knowledge (McGuinness and van Harmelen 2007). This ontology may serve as the basis for further knowledge sharing, refining, and reusing. This ontology explicitly describes collaboration related activity, people, resources, and the relationships in between.

derlying knowledge representation technique. The newly discovered information has to be passed to model refiner to modify accordingly the underlying CM, repository and KM services.

Special attention has to be dedicated to the problem of context uncertainty. There is always uncertainty in context due to the complexity of reality and limitation of sensor (AmI) technologies. Especially innovation processes are related to numerous uncertainties. Various mechanisms for checking reliability of the extracted context (applying, *e.g.*, statistical and reasoning and fuzzy approaches) can be applied. The problem of smooth context extraction in a combined physical and virtual collaboration has to be analyzed.

Ambience sensing services. The main objective of these services is to collect measured data from different AmI systems (and other classical sources) and transform it into knowledge, which can be re-used for context extraction. The services comprehend several parts. AmI information will be collected from the systems in various areas where collaborative work on innovation is taking place and transformed into knowledge.

These AmI systems which may be applied to support innovation processes specifically in manufacturing industry can be grouped into:

1. AmI addressing interaction between operator and machines
2. AmI based sensor networks and systems embedded in equipment monitoring and automation systems
3. System integrated in shop-floor ambience

The AmI data to be collected typically come from:

- Man-machine interfaces such as wearable context aware terminal for maintenance personal and speech recognition systems
- Smart tags and miniaturized information systems
- Cameras, intelligent miniaturized optic sensors, capable of acquiring, storing, processing and transmitting images and video data

Management of social interactions services (set of services supporting collaboration), should generate "raw" information/content on collaborative work in the most appropriate form for future reuse (*i.e.,* for context extraction) and solve critical issues of handling different collaboration patterns, privacy, and Intellectual Property Rights (IPR) issues. Such services include (see Chap. 5):

- Team composition (adjustable to different collaboration patterns)
- Collaboration traceability (easily adjustable to different specific needs/constraints in an EE)
- Services to obtain feedback on knowledge use (collecting feedback from users and/or automatic assessment of knowledge use, *e.g.,* statistical assessment of knowledge resources utilizations, *etc.*)

The critical problem to be solved is how to monitor and document collaborative work without requiring people to document explicitly their activities (automatically) observing privacy and IPR issues, where again the extraction of current *context* may be of key support.

KM services. The set of services for knowledge provision and acquisition (taking into account, *e.g.,* different expertise of teams involved in collaboration) include services for:

- Knowledge/information search using the identified current context
- Knowledge offer for decision making (adapted to the specific context, *e.g.,* user's expertise)
- Human resource discovery (*e.g.,* discovery of experts within an EE available for solving the current problems within innovation tasks and most "adequate" for the actual context regarding time/space constraints, *etc.*)
- Problem solving/reasoning (case base reasoning – CBR)
- Acquiring knowledge from teams (*e.g.,* support in documenting the work done by context-driven provision of knowledge collected within the collaborative situation)

For example, the identified current context may better shape searching space or update the weighting of criteria for similarities in CBR, when searching for similar problems or innovations. These context-enhanced KM services will use dynamically tagged knowledge resources (see above) and/or "similarity" between the current and past stored contexts. The challenge is to effectively take into account *dynamically changing context* in real-time.

Model refiner. The model refiner controls the outer loop of the envisioned system, which is the "high-level" dynamic part of the architecture and which is fundamental for gathering model related knowledge from human users (*via* management of social interactions services and ambience sensing services) and the context provider. This knowledge can then be used for updating context model, the collaborative knowledge repository and the KM services.[21]

The updating of the context model includes automatic adding of new concepts/relations in the existing ontology. The new concepts and relations to be added are extracted from "raw" information delivered by the users, AmI sensors and by MSI services. After updating the context model, the repository must be updated as well.

---

[21] The updating of the collaborative knowledge repository is more complex, because all the existing stored instances must be enriched once the model has been refined. The KM services are then refined in the outer loop, reflecting the changes made in CM and repository. Machine learning technologies can be adopted to refine and enrich the CM. The key problem is how to automatically extract best practice collaboration/activity pattern for different collaboration situations for knowledge reuse.

6.8 Axiomatic Design

An interesting theory for product design has been developed by Dr. Suh Nam Pyo at the renowned Massachusetts Institute of Technology (MIT) in the early 1990s (Suh 1990, 2001, 2005). The theory has been denominated *Axiomatic Design* (AD).

AD offers a very simple definition of design as: "The interplay between what we want to achieve and how we want to achieve it" (Suh 1990) and claims that it enhances creativity (Shinya and Kawassaki 1994) by eliminating bad ideas from the beginning thus allowing design people to focus on promising ones. As Suh states, cited by Gould (2000): "The goal of AD is to make human designers more creative, reduce the random search process, minimize the iterative trial-and-error process, and determine the best design among those proposed."

AD is a system that applies to any kind of design activity, including planning, organization structuring, *etc.,* although it is more fitted to engineering design problems (Leonard and Suh 1994) being applicable to a wide range of fields and products: consumer goods, software development, business processes, *etc.* Some examples can be found in the literature (Yücel and Aktas 2008; Peliks 2003; Brown 2000; Reynal and Cochran 1996). AD has been shown to result in shorter design times, documented design decisions, and ultimately better designs.

AD aims to establish a scientific basis for design based on an axiomatic and algorithmic approach. It takes its name on the idea of "axioms" as basic truths that can't be demonstrated but for which there are no known counter-examples or exceptions. It is also true that axioms are the basis of many scientific and technological fields from Newton's theories to Einstein's achievements.

In consequence, AD postulates two basic axioms:

1. The independence axiom: *"Maintain the independence of the functional requirements (FR)"*
2. The information axiom: *"Minimize the information content of the design"*

Designs complying with both axioms are considered as the best possible designs.

AD postulates a systematic scientific process guiding designers to cross through four design domains: customers, functional, physical, and process (as can be seen in Table 6.3) by linking:

Table 6.3 Four domains of axiomatic design.

Domain	Customers	Functional	Physical	Process
Vectors	Customers' attributes	Functional requirements	Design parameter	Process variables
Acronym	CA	FR	DP	PV

1. Customer attributes (CA) from customer domain to functional requirements (FR) in functional domain
2. FR to design parameters (DP) in physical domain
3. DP to process variables (PV) in process domain

Ideally AD should equal FR and DP maintaining each FR independent from the other (first axiom) and minimize the values in the relationship matrix so reducing the amount of information in the design of each part (second axiom).

Next, "the fun begins" (Gould 2000) when designers decompose the design and establish a hierarchy of FR and DP and begin "zigzagging" between FR ("what") and DP ("how") through the functional and physical domains.

Once the design is settled and the hierarchy levels organized, FR and DP are related through the use of an algorithmic matrix denominated the design matrix A:

$$\{FR\} = [A] \times \{DP\} \quad (6.1)$$

where ideally matrix A should be a square one (equal number of rows and columns) with all values outside the diagonal being "0" so meaning that

$$N^\circ \text{ of FR} = N^\circ \text{ of DP} \quad (6.2)$$

and each FR is related only to a DP. This is known as "*Uncoupled Design.*"

If the matrix A is triangular, then half the values up or down from the diagonal are also "0" and then the design is named: "*Decoupled Design*" in which case the same formula 6.2 is applicable but there exists more than one relationship. *FR are not independent.*

In the remaining cases there exists a *coupled design* meaning that each FR is related to several DP

$$N^\circ \text{ of FR} < N^\circ \text{ of DP} \quad (6.3)$$

From the above-mentioned two axioms, one can derive a set of corollaries defined as "an inference derived from axioms or propositions" proposing simple basic ideas that designers should always have in mind (Suh 1990) such as:

- *Decoupling or separating design elements.* Basically FR have to be decoupled (first axiom) but this does not necessarily implies that physical elements of the design should be separated
- *Minimize FR.* The lower number of FR the simpler the design will result
- *Integrate parts.* The less number of parts the better
- *Standardization.* Standard parts tend to satisfy the design axioms and reduce the information content. They should be used as much as possible
- *Symmetry.* Whenever it is possible to use symmetry, the information content will be reduced

- *Large tolerances.* Using the larger possible tolerances the information content will be reduced
- *Uncouple and minimize information.* The designer should look to reducing the information content and minimizing the interdependence among FR

6.8.1 Axiomatic Product Development Life Cycle

In 2005 Bulent Gumus took a step forward on Suh's theory and developed the *Axiomatic Product Development Life cycle (APDL)* Model from his PhD thesis in Texas Tech University (Gumus 2005). APDL guides the development effort of a transdisciplinary product development team within the extended enterprise (EE) as well as supports them to capture, use and maintain the product knowledge (Gumus *et al.* 2008).

APDL adds a new domain to the previous four in AD: *"The Test domain"*, and four new vectors related to it as can be seen in Table 6.4 (new ones in italics).

Table 6.4 APDL domains and vectors

Domain	Vector	Acronym
Customers	Customer's attribute	CA
Functional	Functional requirement	FR
	Input constraint	*IC*
Test	*Functional test cases*	*FTC*
	Component test cases	*CTC*
Physical	*System component*	*SC*
	Design parameter	DP
Process	Process variable	PV

- *System components* (SC). SC give form to the design solutions complying with the DP. The system components follow a hierarchy that actually represents the system architecture or the product tree explosion
- *Constraints* (IC). IC are boundaries to the design coming either from requirements (CA) not only from customers but for many other sources as regulations, laws, corporation normative, *etc.* or from the limitations of the system (SC) itself on the first steps of the design
- *Test domain.* It covers:
  - Components Test Cases (CTC) for each individual component
  - Functional Test Cases (FTC) as functional testing of the whole product in order to verify its accuracy to the performance of the required functions (Gumus and Ertas 2004).

6.8.2 Similarities and Differences of AD with Other Design Methods

The presented methodology has certain similarities and some new aspects with respect to the other design methods (see Sect. 1.4).

QFD. AD method has a similarity to the QFD process in two senses:

- Both are based on the use of matrices whereas QFD matrices are conceptual and intuitive while AD ones are mathematical matrices with real figures.
- Both link domains. Four in AD: customer, functional, physical, and process; several in QFD which allows jumping among domains through an open user-configurable roadmap of matrices.

As Gould points out (Gould 2000), QFD is more an intuitive method and does not use mathematical relationship as AD does in the design matrix. As discussed in Sect. 3.3.1, QFD macro flow goes from customer requirements to the process quality characteristics through parts decomposition and process specifications, as can be seen in the already known Fig. 6.9, but the four QFD horizontal deployments (Fig. 6.10) open a wide range of possibilities of following different paths through the choice of diverse matrices.

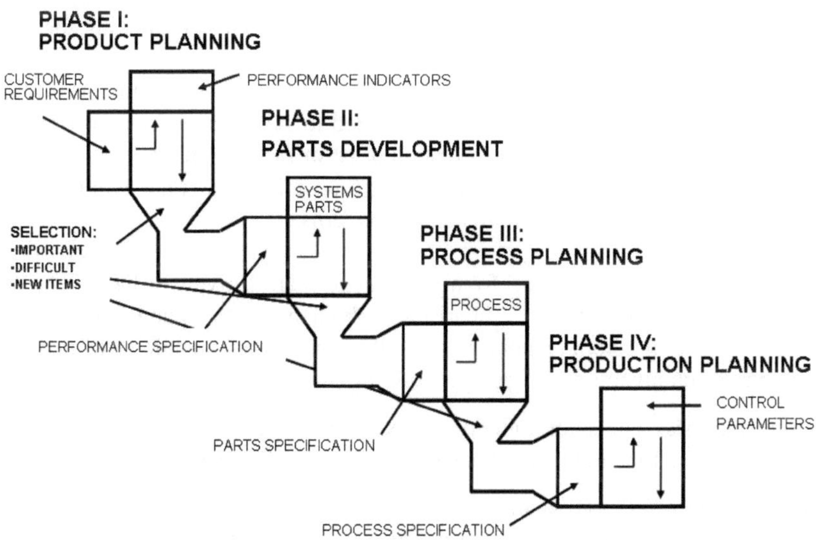

Fig. 6.9 QFD macro flow

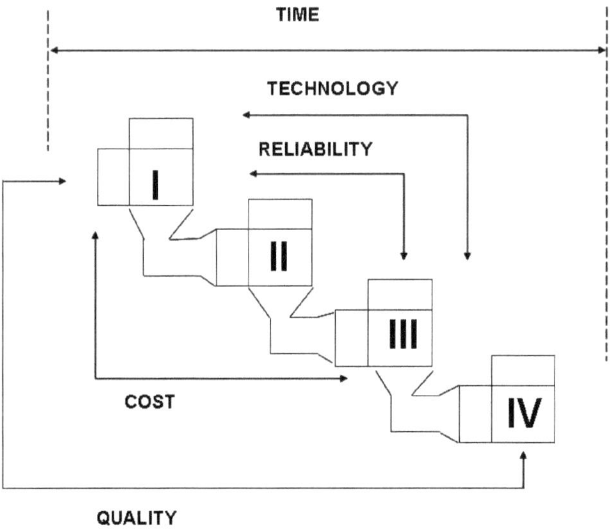

Fig. 6.10 QFD horizontal deployments

TRIZ. Theory for Innovative Problem Solving (TRIZ) also presents some similarities with AD (Mann 1999):

- Basic TRIZ's idea of eliminating contradictions is very close to looking for total independence in FR
- TRIZ's ideality principle of fulfilling the function at the minimum possible costs, ideally without system, is closely related to AD's second axiom (minimizing information) and the ideal situation of N° of FR = N° of DP
- TRIZ also bases on an axiom: *evolution trends* postulating that the evolution of technology is ruled by objective trends and patterns

Kai Yang and Hongwei Zhang from the Wayne State University have done a very good and comprehensive comparison of both methods (Yang and Zhang 2000) to conclude that both techniques are fully compatibles though TRIZ makes more emphases on invention and problem solving (through invention).

Function analysis. Tools for function analysis are a good help to achieve the function decomposition and hierarchical structure in AD (see Sect. 1.4.6).

Taguchi's techniques. AD is considered to be a step beyond Taguchi's methodology in the sense that both intend to produce a design "good from the starting point" and immune to uncontrollable variations (robust design). Nevertheless, Taguchi's techniques deal with design parameters (DP) but does not consider their relationship to the functional requirements (FR) coming from customers (CA) as does AD (see Sect. 1.4.8).

Glossary[1]

Agent An agent can be defined as a computational entity acting on behalf of other entities in an autonomous fashion, performing its actions with some level of proactivity and/or reactiveness, exhibiting some level of the key attributes of learning, co-operation, and mobility. Several agent technologies are operated mainly in the telecommunications realm. They fall into two main categories, *i.e.*, distributed agent technology and mobile agent technology.

Architecture An (ICT) architecture defines the functional capabilities of the parts of a system and the specifications of the relationships between these parts. A layered architecture is based on layers of functional operations. Any layer delivers services to an upper layer and consumes services from underlying layers.

Case-based reasoning It is the process/method of solving new problems based on the solutions of similar past problems. Case-based reasoning is a prominent kind of analogy making.

Choreography Web Service Choreography (WS-Choreography) is a specification by the W3C defining a XML-based business process modelling language that describes collaboration protocols of cooperating Web service participants, in which services act as peers, and interactions may be long-lived. WS-Choreography leverages the power of Web services to allow entities to create business processes that mirror today's dynamic and ever-changing business needs. Organizations can expose their application software and resources as Web services so that others can dynamically find and use them in their business processes. Creating a business process requires not only a clear definition of collaboration patterns of all its components but also a way of depicting standard B2B interactions.

Collaboration Collaboration means working together with shared dynamic goals to achieve collective results that are of benefit to all parties involved. It im-

---

[1] Several terms in the list are overtaken and partly adapted from the list collected by Jean-Pierre Briffaut, GET Institut National des Télécommunications.
http://ec.europa.eu.information_society/activities/atwork/collaboration_at_work/glossary/index_en.htm

plies a higher degree of commitment, mutual trust, sense of belonging, and common interest than cooperation.

Collaborative networks Web connected networks of people working in a collaborative mode on a common task. It implies a conjunction of new working collaborative paradigms and the support of Web-based ICT tools.

Context Context refers to the parts surrounding an object or a subject under consideration. It defines ambient conditions and gives meaning to the situation of an entity.

Context-awareness Context-awareness is the ability to use any piece of context information to contribute to the environmental situation of an entity (person, place, object), thus rendering human-machine interaction more personalized and efficient (discovery and provision of context-aware services). Such context information is delivered by physical and/or logical sensors.

Cooperativity Cooperativity is a neologism characterizing the capability to cooperate.

EDI (Electronic Data Interchange) EDI (Electronic Data Interchange) is an asynchronous exchange of data without human intervention between two application programs run in different computing systems linked by a telecommunication network.

Extended Enterprise It refers to the new working paradigm in which the companies have to interact closely with other companies in the product value chain. The conjoint of enterprises working together in the same product development and manufacturing is named "Extended Enterprise" expanding the enterprise boundaries outside one unique organization.

Extended product Extended product is more than a physical tangible product. Any product has to fulfil a series of customer's requirements intangible and difficult to evaluate. The "extended product" forces a change in the producer's mind from "selling a product" to "fulfilling a service" for which the product itself is just a means.

Groupware Groupware is technology designed to communicate, cooperate, coordinate, solve problems, compete, or negotiate. Typical groupware applications are e-mail, newsgroup, shared whiteboards, decision support systems, multi-player games, *etc.*

Interoperability Traditionally the term interoperability was used in the data processing arena. It is the ability to operate software and exchange data in a heterogeneous network, the best example of which is a large network made up of local area networks. Interoperability testing is an important application of this concept. It covers the integration of a software entity with other software entities to ensure that this entity conforms to alleged standards. Communication between entities without hardware modifications and with easy-made software configuration is the metrics of good interoperability. In collaborative environments the concept of user interoperability is introduced to describe the ability to access and share knowledge between user communities, irrespective of their physical locations, interaction devices, collaborative models, and tools.

Lean manufacturing It is the American denomination for the "Toyota Production System" (TPS). It can be defined as: "Producing just what is needed, at the moment when it is needed, and at the point where it is going to be used" being its main paradigm the complete elimination of waste (anything not adding value to the user). "*What is needed*" should start from the ultimate external consumer (what is actually committed) and pull down the whole value chain of the product.

Legacy systems A legacy system is an (old) computer system or application program that continues to be used because the user (typically an organization) does not want to replace or redesign it.

Model-based reasoning Model-based reasoning refers to an inference method used in reasoning (*e.g.,* expert systems) based on a model of the physical world.

Orchestration Orchestration describes the automated arrangement, coordination, and management of complex computer systems, middleware, and services. It also relates to the process of coordinating an exchange of information through Web service interactions.

OSI (Open System Interconnection) OSI is a reference model for a seven-layer network architecture used for the definition of network protocol standards enabling all OSI-compliant computers or devices to communicate with each other.

Pattern A pattern is in fact a recurring set or sequence of facts or events observed in a given context or domain. An example thereof is a sequence of actions taken by people to solve a given problem (best practices).

Platform In ICT a platform describes some sort of hardware architecture or software framework (including application frameworks) that allows software to run. Typical platforms include a computer's architecture, operating system, programming languages, and related runtime libraries or graphical user interface.

Protocol A protocol is a formally specified set of conventions governing the format and control of inputs and outputs (messages) between two communicating entities (systems, programs...).

Rule-based reasoning Rule-based reasoning is a particular type of reasoning which uses "if-then-else" rule statements.

Scalability In telecommunications and software engineering, scalability is a desirable property of a system, a network, or a process, which indicates its ability either to handle growing amounts of work in a graceful manner, or to be readily enlarged. For example, it can refer to the capability of a system to increase total throughput under an increased load when resources (typically hardware) are added. An analogous meaning is implied when the word is used in a commercial context, where scalability of a company implies that the underlying business model offers the potential for economic growth within the company.

Server In computer networks a server is a unit at a node that provides a specific service for network users, *e.g.,* a data server stores user files, a data processing server offers computing capacity, a remote access server provides access to a LAN for distant users, *etc.*

Service In the context of enterprise architecture, service-orientation, and service-oriented architecture, the term service refers to a discretely defined set of contiguous and autonomous business or technical functionality. OASIS (organization) defines service as "a mechanism to enable access to one or more capabilities, where the access is provided using a prescribed interface and is exercised consistent with constraints and policies as specified by the service description."

SOA (Service-Oriented Architecture) In ICT, service-oriented architecture (SOA) provides methods for systems development and integration where systems group functionality around business processes and package these as interoperable services. SOA also describes ICT infrastructure which allows different applications to exchange data with one another as they participate in business processes. Service-orientation aims at a loose coupling of services with operating systems, programming languages, and other technologies which underlie applications.

SOAP (Simple Object Access Protocol) SOAP is a lightweight XML-based messaging protocol used to encode data in Web service request and response messages before sending them over a network.

UDDI (Universal Description, Discovery and Integration) UDDI is a Web-based directory service that gives businesses and organizations a uniform way to publish their services, discover other companies' services, and understand the procedures required to conduct business with a specific company.

Virtual enterprise A virtual enterprise is a temporary alliance of enterprises that come together to share skills or core competencies and resources in order to better respond to business opportunities, and whose cooperation is supported by computer networks. It is a manifestation of collaborative networks and a particular case of virtual organization.

Virtual reality (VR) Virtual reality is a technology which allows a user to interact with a computer-simulated environment, be it a real or imagined one. Most current virtual reality environments are primarily visual experiences, displayed either on a computer screen or through special or stereoscopic displays, but some simulations include additional sensory information, such as sound through speakers or headphones.

Total Quality (Management) The terminology "total quality" emerged in the early 1970s from Japan with the aim of making people understand that the quality concept has to surpass the boundaries of the traditional quality department, caring about controlling product quality (basically through inspection), should be expanded to the whole organization so becoming "total quality". Within this new framework, product quality becomes responsibility of everybody/everyone in the company not only the quality department and the "total quality" has to be managed, *i.e.,* "total quality management".

Web Service A Web service is defined by the W3C as "a software system designed to support interoperable machine-to-machine interaction over a network." Web services are frequently just Web APIs that can be accessed over a network, such as the Internet, and executed on a remote system hosting the requested services. The W3C Web service definition encompasses many different systems, but in common usage the term refers to clients and servers that communicate over the HTTP protocol used on the Web. Another definition by W3C is "a Web service is a software application identified by a URI (Uniform Resource Identifier) whose interfaces and bindings are capable of being defined, described and discovered as XML artefacts. A Web service support direct interactions with other software agents using XML-based messages exchanged *via* Internet-based protocol." Service providers communicate to the UDDI registry their intents to offer services in an appropriate language (WSDL Web Services Description Language). When a requester wants to consume a service, it searches the UDDI registry to find a service provider. If the registry finds a match, it returns the WSDL description to be interpreted by the client. The client then formats a request and forwards it to the service provider which returns a response.

eXtensible Markup Language (XML) XML is a standard for creating markup languages which describe the structure of data.

References

Ackermann J and Mueller E. (2007) Modeling, planning and designing of logistics structures of regional competence-cell-based networks with structure types. Robotics and Computer-Integrated Manufacturing Vol 23, 601–607

AFNOR-AFCIQ (1981) Guide des Coûts Relatifs á la Qualité. Afnor, France

Afsarmanesh H (2005) D21.1 Characterization of Key Components, Features, and Operating Principles of the Virtual Breeding Environment. Public deliverable of the ECOLEAD project (European Collaborative Networked Organizations Leadership Initiative) http://ecolead.vtt.fi/

AIM (2005) Acceleration of Innovative Ideas to Market. Final Public Report http://www.atb-bremen.de/projects/aim/index.html

Akao Y (1990) Quality Function Deployment. Integrating Customer Requirements into Product Design. Productivity Press. USA

Akao Y (1991) Hoshin Kanri. Policy Deployment for Successful TQM. Productivity Press. USA

Akiyama K (1991) Function Analysis: Systematic Improvement of Quality and Performance. Productivity Press, USA

Altshuller G (1988) Creativity as an Exact Science. Gordon and Breach, USA

Altshuller G (1992) And suddenly the inventor appeared. Technical Innovation Center Inc, Massachusetts, USA

Altshuller G (1996) And Suddenly the Inventor Appeared. Technical Innovation Center Inc, USA

Altshuller G (1997) 40 Principles. TRIZ Keys to Technical Innovation. Technical Innovation Center Inc, USA

Andrews P (2006) Five barriers to innovation: Key questions and answers. IBM Executive Technology Report
http://www.935.ibm.com/services/us/index.wss/executietech/gbs/a1025989?cntxt=a1005266

Arciszewski T and Zlotin B (1998) Ideation/TRIZ: Innovation Key to Competitive Advantage and Growth. Innovation International Inc,
http://www.ideationtriz.com/paper_ITRIZ_Innovation_Key.asp

Arpírez JC, Corcho O, Fernández-López M and Gómez-Pérez A (2001) WebODE: a Scalable Workbench for Ontological Engineering. First International Conference on Knowledge Capture (KCAP01). Victoria, Canada. October.

Ashburn A (1977) Toyota's famous Ohno System (a reprinted version of the article in the "American Machinist, July 1977"). In Monden Y (ed) (1986) Applying Just In Time: The American/ Japanese Experience. Ind Eng and Manag Press. IIE

ASSIST (2006) Knowledge-based intelligent design assistant. http://www.labein.es/assist

Bair J (1997) Knowledge Management Leverages Engineering at Chrysler. Gartner Group Research Note, January

Barrentine LB (1999) An Introduction to Design of Experiments: A Simplified Approach. ASQ Quality Press, Milwaukee, WI, USA

Bekhti S, Matta N (2002) Traceability and knowledge modelling. COOP2002

Bhushan N (2007) Set Based Concurrent Engineering (SBCE) and TRIZ – A Framework for Global Product Development. TRIZCON2007 (9th Annual Conference of the Altshuller Institute for TRIZ Studies) Louisville, KY, USA

Biuk-Aghai, R P (2004) Patterns of Virtual Collaboration in Online Collaboration Systems. Proceedings of the IASTED International Conference Knowledge Sharing and Collaborative Engineering, St. Thomas, US Virgin Islands, 22–24 November, pp. 57–62, ACTA Press

Bower JL (1970) Managing the resource allocation process. Harvard Business School Press. Boston, MA

Box GEP (1988) Signal to noise ratios, performance criteria, and transformations. Technometrics 30: 1–17

Box GEP, Hunter WG and Hunter JS (1978). Statistics for experimenters. Wiley, Nueva York

Brassard M (1989) The Memory Jogger +. GOAL/QPC

Braungart M and McDonough B (1998) The NEXT Industrial Revolution. Atlantic Mon Mag 282/4, October

Brookson C and Zumerle D (2006) ETSI White Paper No. 1; Security for ICT – the Work of ETSI. Sophia Antipolis, January

Brown CA (2000) Axiomatic Design of Chaotic Components of Surface Textures. Proc of ICAD2000 First International Conference on Axiomatic Design. Cambridge MA June. Institute for Axiomatic Design, Cambridge, Derrick Tate, ed., 106–111 http://www.axiomaticdesign.com/technology/icad/icad2000/icad2000_008.pdf

Budzik J, Hammond K (1999) Watson: anticipating and contextualizing information needs, Proc of the 62nd Annual Meeting of the American Society for Information Science

Burgelman RA (1983a) A model of the interaction of strategic behaviour, corporate context, and the concept of strategy. Academy of Management Review, 8: 6170

Burgelman RA (1983b) A Process Model of Internal Corporate Venturing in the Diversified Major Firm. Adm Sci Q 28/2: 223–244

Burgelman RA (1994) Fading Memories: A process theory of strategic business exit in dynamic environments. Administrative Science Quarterly, 39/1: 24–56

Buzan T (2003) The Mind Map Book: Radiant Thinking – Major Evolution in Human Thought BBC Active (paperback 2003) ISBN 0-563-48701-1

Byrne DM and Taguchi S (1987) The Taguchi approach to parameter design. Qual Prog. Dec

Campos A, Stokic S, and Neves-Silva R (2004) Integrated Approach for Innovation and Problem Solving in Dynamic Virtual Enterprises. INDIN 04, Berlin, 24–26 June

Campos A, Pina P and Neves-Silva R (2006) Supporting Distributed Collaborative Work in Manufacturing Industry. Proc of the Second International Conference on Collaborative Computing: Networking, Applications and Worksharing, CollaborateCom

Campos A, Stokic D, Correia A, Agirre-Ibarbia J, Pina P and Neves-Silva R (2007) Supporting eCollaboration in Manufacturing Industry. Proc of the Seventh eChallenges e-2007 Conference and Exhibition

Castro Reis D, Golgher PB, Da Silva AS and Laender AF (2004) Automatic Web News Extraction Using Tree Edit Distance. Proc of the International WWW Conference, New York, USA

Charkiewicz E, Bennekom S and Young A (2001) Transitions to sustainable production and consumption: concepts, policies, and actions. The Hague: Tools for Transition

Chaudhari A and Patil V (2002) Future Trends in Collaborative Product Data Management Systems. Tata Consultancy Services

Chesbrough H W and Teece D J (1996) When is Virtual Virtuous? Organizing for Innovation. Harvard Business Review

Christensen C M and Bower J L (1996). Customer power, strategic investment, and the failure of leading firms. Strategic Management Journal, 17/3: 197–218

Christensen C (1997) Innovator's dilemma: When new technologies cause great firms to fail. Boston, MA: Harvard Business School Press.

Christensen C, Johnson CW and Horn MB (2008) Disrupting Class: How Disruptive Innovation Will Change the Way the World Learns. McGraw-Hill Professional, New York, London

Churchill E, Snowdon D and Munro A (2001) Collaborative Virtual Environments, Digital Places and Spaces for Interaction. Springer-Verlag, Berlin

Clausing D and Simpson B.H (1990) Quality by Design. Quality Progress 23/1: 41–44

Coenen L and Díaz López F (2008) Comparing systemic approaches to innovation for sustainability and competitiveness. DIME International Conference, France

Cohen M and Murphy J (2001) Exploring sustainable consumption environmental policy and the social sciences. Oxford Pergamon

Collins J (2001) Good to Great: Why Some Companies Make the Leap... and Others Don't. Random House Business, London

Correia A, Grama C, Stokic D, Campos A, Neves-Silva R and Aguirre J (2008a) An Architecture for Collaborative Working Environments in Manufacturing Industry. Proc of the 15th Annual European Concurrent Engineering Conference, ECEC 2008

Correia A, Stokic D, Neves-Silva R, Campos A, Agirre J (2008b) Collaborative Environment for Intelligent Monitoring in Manufacturing Industry. Proc of the 14th International Conference on Concurrent Enterprising, ICE2008

Correia AT, Grama C, Stokic D *et al.* (2008c) An Architecture for CWE in Manufacturing Industry. Proc of the16th European Concurrent Engineering Conference, Porto

Cosulschi M, Giurca A, Udrescu B, Constantinescu N and Gabroveanu M (2006) HTML Pattern Generator – Automatic Data Extraction from Web Pages. Proc of 8th International Symposium on Symbolic and Numeric Algorithms for Scientific Computing, SYNASC '06, Timisoara, Romania

CuteLoop (2008) State of the-Art Analysis. Public Report, CuteLoop Research Project, http://www.cuteloop.eu

Dailey K (2004) The FMEA Pocket Handbook. DW Publishing Co

De Bono, E (1967) The Use of Lateral Thinking. Jonathan Cape, London

De Bono E (1985) Six Thinking Hats. Key Porter, Toronto

De Laat P B (1999) Systemic innovation and the virtues of going virtual: The case of the digital video disc. Technology Analysis & Strategic Management, 11/2: 159–180

De Vries S and Kommers P (2004). Online knowledge communities: future trends and research issues, Int J of Web Based Communities 1(1): 115–123

Decker S (2006) Towards a Reference Architecture for Collaborative Environments, ECOSPACE Newsletter, No.1, (http://www.ip-ecospace.org), October, 2006

Deming WE (1975) On probability as a basis for action. The American Statistician, 29(4): 146–152

Deming WE (1986). Out of the Crisis. MIT Press

Dornier (1992) A&R Control Development Methodology Definition Report. Doc. Nr. CT2/CDR/DO Issue 2.0 Sept

Drira K, Martelli A, Villemur T (2001) Cooperative Environments for Distributed Systems Engineering. The Distributed Systems Environment Report (Lecture Notes in Computer Science Vol. 2236). Springer-Verlag, Berlin, Heidelberg

Edquist, C (1999) Innovation Policy – A Systemic Approach. DRUID's Innovation Systems Conference, Aalborg http://www.druid.dk/conferences/summer1999/conf-papers/edquist.pdf

Edquist, C (2001) Innovation Policy – A Systemic Approach. Major Socio-Economic Trends and European Innovation Policy. Oxford University Press, Oxford, UK

e-Mult (2006) System Concept. Public report of the E-Mult Project [IST-2004-027212 (2006–2008)]. http://www.e-mult.eu/

Epstein, S (1994) Integration of the cognitive and the psychodynamic unconscious. American Psychologist 49, 709–724

Ernst & Young and SPRU (University of Sussex) (1998) Integrated Product Policy Executive summary. European Commission DG XI http://ec.europa.eu/environment/ipp/pdf/ippsum.pdf

EUROBUNT (1995a) Eurobunt Toolkit. Strategic Innovation. Eurobunt, Norway

EUROBUNT (1995b) The European Handbook of Management Consultancy. Strategic Innovation: An European Approach to Management Consultancy

European Commission (1995) Green Book on Innovation. COM (95) 0688, Dec http://europa.eu.int/comm/off/pdf/COMM(95)688.pdf

European Commission-DGXIII (1995) Value Management Handbook

European Commission (2001) Green Paper on Integrated Product Policy COM(2001)68 http://ec.europa.eu/environment/ipp/2001developments.htm

Expert Group (2004) New Working Environments, Next Generation Collaborative Working Environments 2005–2010. DG Information Society http://europa.eu.int/information_society/activities/atwork/work_paradigms/experts_group/documents/next_generation_collab_environments_report.pdf

Expert Group (2005) The Digital Ecosystems Research Vision: 2010 and Beyond. July 2005, European Commission Expert Group

Expert Group (2006) Towards Activity-oriented Collaborative Working Environments. A Research Roadmap 2007–2020. DG Information Society. European Commission. Rapporteurs: Nikolay Mehandjiev and Dragan Stokic

Farquhar A, Fikes R and Rice J (1997) The Ontolingua Server: a Tool for Collaborative Ontology Construction; Intl J of Hum-Comp Studies 46

Feigenbaum AV (1991) Total Quality Control. McGraw-Hill

Fernández López M, Gómez-Pérez A, Pazos Sierra J and Pazos Sierra A (1999) Building a Chemical Ontology Using Methontology and the Ontology Design Environment, IEEE Intelligent Systems, 14, 1, 37–46

Fiore C (2005) Accelerated Product Development. Productivity Press. USA

Ford DF and Sobek II DK (2005) Adapting Real Options to New Product Development by Modelling the Second Toyota Paradox. IEEE Transactions on Engineering Management. 52/2: 175–185

Gartner B and Wille R (1999) Formal Concept Analysis: Mathematical Foundations, Springer-Verlag

Gertman, DL and Blackman, HS (2001). Human reliability and safety analysis data handbook. Wiley

Gilbreth, F (1911) Motion Study. NY D Van Nostrand Co

Gilbreth, F & Gilbreth L (1917) Applied Motion Study. NY Sturgis & Walton Co

Gilbert C G (2002) Beyond Resource Allocation: Towards a Process Model of Response to Disruptive Change. Harvard Business School Working Paper #03

Goldratt E (1990) The Haystack Syndrome – Sifting Information Out of the Data Ocean. North River Press (USA)

Goldratt E (1996) The Goal. A process of ongoing improvement. North River Press (USA)

Goldratt E (1997) Critical Chain. North River Press (USA)

Gómez-Pérez A, Fernández-López M and Corcho O (2004) Ontological Engineering. Springer-Verlag, London, Berlin

Gorostiza A, Sorli M, Stokic D and Scholze S (2005) Knowledge Based Intelligent Design Assistant. Pro of ICE 2005 Munich

Gould L (2000) Building Better Vehicles via Axiomatic Design. Automotive Design and Production. June http://www.autofieldguide.com/articles/060001.html

Grand S, von Krogh G, Leonard D and Swap W (2004) Resource allocation beyond firm boundaries: A multi-level model for open source innovation. Long Range Planning, 37/6: 591–610

Groocock JM (1986) The Chain of Quality. John Wiley & Sons

Große Hovest G, Grama C, Urošević L and Stokić D (2008) Collaborative Environment for Virtual Collaborative Networks of ELV Recycling SME. Proc of ICE2008, Lisbon

Guarino N (2002) Evaluating Ontological Decisions with OntoClean. Comp Sci & Eng 45: 61–65

Gumus B (2005) Axiomatic Product Development Lifecycle (APDL) Model. PhD Dissertation, TTU http://etd.lib.ttu.edu/theses/available/etd11282005154139/unrestricted/Gumus_Bulent_Diss.pdf
Gumus B and Ertas A (2004) Requirements Management and Axiomatic Design. J Int Des and Process Sci. 8/4: 19–31. Dec 2004
Gumus B, Ertas A, Tate D, and Cicek I (2008) Transdisciplinary Product Development Lifecycle Framework and its Application to an Avionics System. J Eng Des. 19/3:185–200, doi: 10.1080/09544820701232436
Halpin JF (1970) Zero Defects. CEAC (Spanish ed)
Haque B and James-Moore M (2004) Applying Lean Thinking to New Product Introduction, J Eng Des Volume 15, No. 2
Harrison A, Pavitt J and Alexander J (2002) The Status of Lean Thinking in UK Lean Aerospace Initiative (UK-LAI) Supply Chains: a Survey. School of Management, Cranfield University
Hauser J R and Clausing D (1988) The House of Quality. Harvard Business Review, May–June
Heinezahll L (1996) Total Quality Leadership. The Future Leaders in Quality Organisations. The 1996 EFQM Learning Edge Conference
Henderson C (2006) Have you performed a Kaizen? N J Inst Health Rec Inf Manag 47/1: 4, 6–8
Henderson R and Clark K (1990) Architectural Innovation: The Reconfiguration of Existing Product Technologies and the Failure of Established Firms. Administrative Science Quarterly 35 (1) Special Issue: Technology, Organizations, and Innovation
Hendryx S (2003) Architecture of Business Modelling. OMG Document
Herbsleb JD and Mockus, A (2003) An Empirical Study of Speed and Communication in Globally-Distributed Software Development, IEEE Transactions on Software Engineering, 29, 3
Hicks CR (1982) Fundamental concepts in the design of experiments. Holt, Rinehart and Winston. New York
Hinkelmann K and Kempthorne O (1994) Design and analysis of experiments. Volume I: Introduction to experimental design. Wiley, New York
Hollanders H (2007) 2006 European Regional Innovation Scoreboard (2006 RIS). European Trend Chart on Innovation
Horváth I and van der Vegte WF (2003) Nucleus-based product conceptualization: principles and formalization. Proc of ICED'03, 14th international conference on engineering design, Stockholm
Horváth I, Vergeest JSM and Rusák Z (2002) Six ingredients of collaborative virtual design environment. International Design Conference, Dubrovnik
Hubbard D (2007) How to Measure Anything: Finding the Value of "Intangibles" in Business. Wiley- Higher Education
Huthwaite B (2007) The Lean Design Solution. Peppertree Press
Imai M (1986) Kaizen: The Key to Japan's Competitive Success. New York, Random House
InAmI (2006) Presentation of InAmI Project and InAmI eCollaboration Public Concept. Deliverable D4 InAmI project (NMP-IST- 2004-016788): Innovative ambient intelligence based services to support life-cycle management of flexible assembly and manufacturing systems, http://www.uninova.pt/inami
Ishikawa K (1985) What is Total Quality Control? The Japanese Way. Prentice Hall.
ISO 12006-2 Building construction (2001) Organization of information about construction works – Framework for classification of information. DIS, Brussels
Jackson B and Ceusters W (2002) A novel approach to semantic indexing combining ontology-based semantic weights and in-document concept co-occurrences. In Baud R, Ruch P. (eds) EFMI Workshop on Natural Language Processing in Biomedical Applications, 8–9 March, 2002, Cyprus, pp. 75–80
Jerrard R, Barnes N and Reid A (2008) Design, Risk and New Product Development in Five Small Creative Companies. J Des 2/1: 21–30

Johansson F (2004) The Medicci Effect. Breakthrough Insights at the Intersection of Ideas, Concepts and Cultures. Harvard Business School Press
Juran JM (1962) Quality Control Handbook. McGraw-Hill
Juran JM (1993) Made in USA: A Renaissance in Quality. M.-Juran Institute, Inc. Harvard Business Review, USA
Juran JM (1995) A History or Managing for Quality. The Evolution, Trends, and Future Directions of Managing for Quality. Editor-in-chief. ASQC Quality Press
Juran JM, Bingham RS, Gryna, FM and Vallhonrat JM (1983) Manual de Control de Calidad. Editorial Reverté, S.A. Spain
Kano N. (1984) Attractive quality and must-be quality. J Jap Soc QC April 39–48
Kano S (2000) Technical Innovations, Standardization and Regional Comparison - a Case Study in Mobile Communications. Telecommunications Policy, 24/4: 305
Karlsson C and Ahlstrom P (1996) The difficult path to Lean product development. J Prod Innov Manag 13: 283–295
Keil T (2002) De-facto standardization through alliances – lessons from Bluetooth. Telecommunications Policy, 26/3-4: 205–213
Kennedy M (2003) Product Development for the Lean Enterprise: Why Toyota's System is Four Times More Productive and how you can implement it. Oaklea Press, Richmond, Va
King B (1989) Better Designs in half the time. GOAL/QPC, Boston
K-NET (2008) Public Concept. K-NET project (ICT-2007-215584): Services for context sensitive enhancing of knowledge in networked enterprises, http://140.203.154.228/K-net/index.jsp
Knight F (1921) Risk, uncertainty and profit. Houghton Mifflin, New York
Know-Construct (2007) Internet Platform for Knowledge-based Customer Needs Management and Collaboration among SMEs in Construction Industry. EU project COLL-CT-2004-50027 Final Public Report, http://www.know-construct.com/index.htm
Kogut B, Metiu A (2001). Open-source software development and distributed innovation. Oxford Review of Economic Policy, 17/2: 248–264
Kohnhauser V (1999) Use of TRIZ in the Development Process, Triz Journal, June
Kotter JP and Heskett JL (1992) Corporate Culture and Performance. Free Press ; New York; Maxwell Macmillan International, Oxford
Kraut R, Edigo C and Galegher J (1990) Patterns of Contact and Communication in Scientific Research Collaboration. In Intellectual Teamwork: Social and Technological Foundations of Cooperative Work Lawrence Erlbaum Associates, Hillsdale New Jersey, pp. 149–171
Kuczynski A, Stokic D and Kirchhoff U (2005) Set-up and Maintenance of Ontologies for Innovation Support in Extended Enterprises. Int J of Adv Manuf Technol, Springer Verlag, London, Berlin
Kume H (1993) Quality Management by ISO 9000 and by TQM. ISO 9000 Forum. EOQ. Helsinki, Finland
Laso-Ballesteros I and Salmelin B (2005) AMI-endowed Collaboration@work, Ambient Intelligence in Practice: Future Perspectives and Applications. Riva G, Vatalaro F, Davide F and Alcañiz M (Eds) IOS Press, 237–264
Leonard B (2001) Funding of New Technology-Based Firms by Commercial Banks in Europe. DIANE Publishing, Collingdale
Leonard DA and Suh NP (1994) Axiomatic Design and Concurrent Engineering. Comp-Aided Des, 26/7: 499–504
Li G, Bracewell RH and Wallace KM (2001) A Knowledge Framework to Support Engineering Design. Proc of 8th ISPE International Conference on Concurrent Engineering: Research and Applications, California, USA
Li WD, Ong SK and Nee AYC (2006) Integrated and Collaborative Product Development, Technologies and Implementation. World Scientific
Liker J (2006) The Toyota Way Fieldbook. New York, McGraw-Hill

Litaudon, M. and Réfabert, A. (1992) Análisis del Valor para la mejora de productos. Gestión 2000. Barcelona

Luque MA and Montoya G (1995) Introducción al Análisis de Valor. IAT (Instituto Andaluz de Tecnología) Sevilla

Mann DL (1999) Axiomatic Design and TRIZ: Compatibility and Contradictions. TRIZ J, June and July

Mascitelli R (2007). The Lean Product Development Guidebook: Everything Your Design Team Needs to Improve Efficiency and Slash Time-to-Market. Technology Perspectives

Maula MVJ, Keil T and Salmenkaita JP (2006) Open innovation in systemic innovation contexts. Oxford University Press, Oxford http://www.openinnovation.net/Book/NewParadigm/Chapters/12.pdf

Mazur G (1993) QFD for Service Industries. From Voice of Customer to Task Deployment. Proc of the 5th Symposium on QFD, Novi (Michigan) USA

MacHale D (2002) Wisdom. Prion, London

McDonough W and Braungart M (2002) Cradle to cradle: remaking the way we make things. North Point, New York

McGuinness D (1998) Ontological Issues for Knowledge-Enhanced Search; Proc of Formal Ontology in Information Systems, June 1998. Also in Frontiers in Artificial Intelligence and Applications, IOSPress, Washington, DC

McGuinness D and van Harmelen F (2007) OWL Web Ontology Language – Overview. W3C Recommendation, www.w3.org/TR/owl-features/, April 10, 2007

Meier D, Tautz C, Traphöner R, Wissen M and Ziegler j (2001). Building Ontologies for Knowledge Management Applications in Group Sessions. K-CAP 2001, 20.10.2001, Victoria, B.C., Canada

Mena E and Illarramendi I (2001) Ontology-Based Query Processing for Global Information Systems. Kluwer Academic Publishers, ISBN 0-7923-7375-8, pp. 215, June 2001

Mendikoa I and Sorli M, (2005) Distributed Product Design and Manufacturing based on KBE. Lecture Notes in Computer Science 3865. Springer-Verlag Berlin Heidelberg, 2006, ISSN 0302-9743 (Paper selected from "Computer Supported Cooperative Work in Design II", Coventry, 2005)

Mendikoa I, Sorli M, Barbero JI and Carrillo A (2005) Knowledge Based Distributed Product Design and Manufacturing. Proc of the Ninth International Conference on Computer Supported Cooperative Work in Design (CSCWD): 679–685. May, Coventry, UK

Mendikoa I, Sorli M, Barbero JI and Carrillo A (2008) Collaborative Product Design and Manufacturing with Inventive Approaches. Int J Prod Res. 46(9): 2333–2344

Merli G (1994). Nueva Estrategia de Aprovisionamiento para la Fabricación: Comakership (Fabricación Asociada). Díaz de Santos, Spain

Mezura-Godoy C and Talbot S (2001) Towards social regulation in computer-supported collaborative work. Seventh International Workshop on Groupware

Miao Y and Haake JM (1998) Supporting Concurrent Design by Integrating Information Sharing and Activity Synchronization. CE'98. Tokyo, Japan, July 15–17

Michalko M (1991) Thinkertoys. Ten Speed Press. Berkeley, California (USA)

Miles LD (1989) Techniques of Value Analysis and Engineering. Wendt Library http://wendt.library.wisc.edu/miles/index.html

Missikoff M, Velardi P and Navigli R (2002) The Usable Ontology: An Environment for Building and Assessing a Domain Ontology, Proc of the Int Semantic Web Conference 2002 (ISWC2002), Sardinia Italy

Mizunu S. (1988) Company-Wide Total Quality Control. Asian Productivity Organization

Mobley RK (1999) Root Cause Failure Analysis. Butterworth-Heinemann

Molli P, Skaf-Molli H, Oster G and Jourdain S (2002) SAMS: Synchronous, asynchronous, multi-synchronous environments. Proc of the Seventh International Conference on CSCW in Design. Int Work Group on CSCW in Des, September

Monden Y (1987) El Sistema de Producción de Toyota. IESE. Spain

Montgomery DC (1991) Design and analysis of experiments. Wiley, NewYork
Morgan J and Liker JK (2006) The Toyota Product Development System: Integrating People, Process and Technology. Taylor & Francis, Inc
Murman E, Allen T and Cutcher-Gershenfeld, J (2002) Lean Enterprise Value: Insights from MIT's Lean Aerospace Initiative. Palgrave
NATO (2007) Advanced Research Workshop on Sustainable Energy Production and Consumption and Environmental Costing. Naples, Italy
Nemiro EJ (2004) Creativity in Virtual Teams: Key Components for Success, Pfeiffer
Ning K, Gong R, Decker A, Chen Y and O'Sullivan D (2007) A Context-Aware Resource Recommendation System for Business Collaboration. EEE2007
Noda T and Bower J L (1996) Strategy making as iterated processes of resource allocation. Strategic Management Journal, 17, 159–192
Noy NF and Musen MA (2000) PROMPT: Algorithm and Tool for Automated Ontology Merging and Alignment. In: Seventeenth National Conference on Artificial Intelligence (AAAI-2000). Austin, TX
Noy N F and Musen MA (2001) Anchor-PROMPT: Using Non-Local Context for Semantic Matching. In: Workshop on Ontologies and Information Sharing at the Seventeenth International Joint Conference on Artificial Intelligence (IJCAI-2001). Seattle, WA, 2001
NWE-EC (2004) Collaboration@Work: The 2004 report on new working environments and practices. Unit F4 – New Working Environments, Directorate-General Information Society and Media. Office for Official Publications of the European Communities, Luxembourg
NWE-EC (2005) Collaboration@Work: The 2005 report on new working environments and practices. Unit F4 – New Working Environments, Directorate-General Information Society and Media Office for Official Publications of the European Communities, Luxembourg
Oakey RP (1994) New Technology-Based Firms in the 1990s. Paul Chapman Publishing. London
Oakland J (2002) Statistical Process Control. Butterworth-Heinemann, Oxford
OECD (1997) OECD programme on sustainable consumption and production. OECD Workshop on Improving the Environmental Performance of Government. Paris: OECD
O'Grady PJ and Kim C (1966) Feature-based design of electronics assemblies. Int. J. of Prod. Research, 34, 5, 1307–1330
Ohno T (1988) Toyota Production System. Productivity Press, USA
Ortiz C (2006) All-out kaizen – A continuous improvement plan delivers change to the production floor... and dollars to the bottom line Industrial Engineer Volume 38, Issue 4, pp. 30–34 APR
Osborn AF (1942) How to Think Up. McGraw-Hill Book Co. New York & London
Osborn AF (1949) Your Creative Power. How to use imagination. Charles Scribner's Sons. New York, London
Osborn AF (1979) Applied Imagination. CEF Press Buffalo, New York
Pahl G, Beitz W (1996) Engineering Design: a systematic approach. (2nd edn.). New York, Springer-Verlag
Papazoglou MP and Van den Heuvel WJ (2007) Service Oriented Architectures: Approaches, Technologies and Research Issues, 2007. 3, s.l.: Springer, 2007, The International Journal on Very Large Data Bases (VLDB), Vol. 16, pp. 389–415
Patel D, Collins R, Vanfleet WM, Calloni BA, Wilding MM, MacLearn L and Luke JA (2002) Deeply Embedded High Assurance (Multiple Independent Levels of Security/Safety) MILS Architecture. November, 13
Pekka P (2000) Human Reliability Analysis methods for probability safety assessment. VTT Publications
Peliks RB (2003) Axiomatic redesign: using Axiomatic Design to improve vehicle performance of the steering and suspension system. Thesis at MIT
Peltz C (2003) Web Service Orchestration and Choreography – a look at WSCI and BPEL4-WS, www.wsj2.com, July

Piatier A (1984) Barriers to Innovation. Commission of the European Communities, Directorate-General Information Market and Innovation. Publisher: F. Pinter

Pihkala T, Ylinenpaa H and Vesalainen (2002) Innovation barriers amongst clusters of European SMEs. Int J of Entrep and Innov Mang 2/6: 520–553

Placzkowski G (2000) World class quality: Using design of experiments to make it happen. Qual Prog 33/10: 108–108

Prasard B (1997) Concurrent Engineering Fundamentals. Prentice Hall International USA

Prinz W (2005) Activity Oriented Collaboration Systems. Collaboration@Work: The 2005 report on new working environments and practices. Office for Official Publications of the European Communities, Luxembourg

Ralli C (2005). Collaboration Reference Architecture. Workshop on eCollaboration in working environments VUB. Brussels, November, 2005.

Rector A, Bechhofer S, Goble C, Horrocks I, Nowlan W and Solomon W (1997) The GRAIL Concept Modelling Language for Medical Terminology. Artificial Intelligence in Medicine, Volume 9

Réfabert A and Litaudon M (1988) La Dynamique de l'analyse de la valeur. Les Éditions d'organisation, Paris

Reynal VA and Cochran DS (1996) Understanding Lean Manufacturing According to Axiomatic Design Principles. The Lean Aircraft Initiative Report Series # RP96-07-28 MIT. November

Rogers JE, Roberts A, Solomon WD, Van der Haring E, Wroe CJ, Zanstra PE, and Rector AL (2001) GALEN Ten Years On: Tasks and Supporting tools Proc of MEDINFO2001, V. Patel *et al.* (Eds) IOS Press: 256–260

Roy, RK (2001) Design of Experiments using the Taguchi Approach. John Wiley & Sons

Roy, RK (2003) Cost Engineering: Why, what and how?. Decision Engineering Report Services, Cranfield University, UK

Rushby J (1981) Design and Verification of Secure Systems ACM Operating Systems Review, 15, 5, Reprint of a paper presented at the 8th ACM Symposium on Operating System Principles, Pacific Grove, California, 12–21

Saderra L (1994) La Calidad Total: el secreto de la industria japonesa. Ed. Pioneer

Salles L and Furtado V (2004) Modelling Multi-Agent Systems for Knowledge Management in the Software Development Process. Second Agent-mediated Knowledge Management Workshop. Valencia, Spain. Proc of AMKM 2004, v. 1

Sarma A (2005) A survey of collaborative tools in software development University of California Irvine, Institute of Software Research, Technical Report, UCI-ISR-05-3, March 2005

Schaffers H, Brodt T, Prinz W and Pallot M (2006) The Future Workspace – Perspectives on Mobile and Collaborative Working http://www.ami-communities.eu/pub/bscw.cgi/d163187/The%20Future%20Workspace.pdf

Seppola R (2004) Social Capital in International Business Networks. Helsinki School of Economics

Shewhart WA (1939) Statistical Method from the Viewpoint of Quality Control. The Graduate School; the Dept. of Agriculture, Washington

Shillito ML and De-Marle D (1992) Value: Its Measurement, Design and Management. Ed. John Wiley & Sons, New York

Shinya S and Kawassaki U (1994) A Study of Creative Design Based on the Axiomatic Design Theory, DE-Vol.68, Design Theory and Methodology-DTM'94 ASTM 1994

Siemens (2002) Idea Management, the 3i Program, Cross-company agreement for the 3i Program. Published by Corporate Personnel (CP G ISA)

Slovic P, Finucane ML, Peters E and MacGregor DG (2004) Risk as Analysis and Risk as Feelings: Some Thoughts about Affect, Reason, Risk, and Rationality. Risk Analysis 24/2: 311–322

Smith A (1998) An Inquiry into the Nature and Causes of the Wealth of Nations. (first published in 1776) Oxford University Press, New York

Soares A L, Silva M, Simões D and Costa R (2005) Know-Construct: Knowledge based community management in the Construction Industry. Proc of the 7th Terminology and Knowledge Engineering Conference; Workshop Working with Terminology and Knowledge Management Systems. Copenhagen: GTW

Sobek II DK (1997) Toyota's Product Development Process. A case study in Fleischer M and Liker J (ed) Concurrent Engineering Effectiveness: Concepts and Methods. Hanson-Gardner Publishers. pp: 461–480

Sobek II DK, Liker LK and Ward AC (1999) Toyota's Principles of Set-Based Concurrent Engineering. Sloan Manag Rev. 40/2:67–83

Sorli M and Gómez A (1994) Integration of Total Quality Methodologies with Simultaneous Engineering Concepts in a Comakership Frame. SITEV 94. International Week of the Automotive Industry. Torino. November 15th

Sorli M and Gutiérrez JA (2002) New Paradigms in product Development. Engineering Design Conference (EDC2002). Imperial College, London

Sorli M and Mañà F (1997) Strategic Product Innovation using QFD and VM. Proc 3rd International Symposium on QFD, Sweden

Sorli M and Ruiz J (1994) QFD. Una Herramienta de Futuro. Labein, Bilbao (Spain)

Sorli M and Stokic D (2008) Supporting Innovation through Knowledge Management in the Extended Enterprise. In Zhao F (ed) Information Technology Entrepreneurship and Innovation. IGI Global: 310–329

Sorli M, Stokic D, Campos A, Sanz A and Lagos MA (2002a). Fostering Innovative Ideas and Accelerating them into the Market. e-Business Conference 2002, Prague, Czech Republic

Sorli M, Gutiérrez JA and Zubiaga R. (2002b) Implications of the Integrated Product Policy (IPP) in the new products design and development. In Hon B (ed) Design and Manufacturing for Sustainable Development. Professional Engineering Publishing, UK

Sorli M, Gutierrez JL and Mazharsolook A (2003) A new Model for Product Development underpinned by an Information-Management System. J of Comp in Comp & Eng. 2003/4: 38–47

Sorli M, Stokic D, Gorostiza A and Campos A (2004) Acceleration of Innovative Ideas to Market. IMS Forum, Como (Italy)

Sorli M, Barbero JI, Mendikoa I and Carrillo A (2005) Distributed Product Design and Manufacturing. Web-enabled collaboration design and manufacture. In Daizhong Su (ed) Literature Review, Research and Development. Nottingham Trent University, UK

Sorli M, Stokic D and Gorostiza A (2006a) Supporting Innovation Management in Extended Enterprise. Proc of the International Conference on Advanced Design and Manufacture. Harbin, China
http://www.chinagrid.net/dvnews/show.aspx?id=1135&cid=11
http://www.admec.ntu.ac.uk/adm2006

Sorli M, Mendikoa I, Pérez J, Soares A, Urosevic L, Moreira J and Corvacho H (2006b) Knowledge-based Collaboration in Construction Industry. 12th Int Conf on Concur Enterp Milan: 26–28, http://www.ice-conference.org

Sorli M, Stokic D, Gorostiza A and Campos A (2007a) Fostering innovation in practice through TRIZ-based CAI tool. Int J Comp Applic Tech (IJCAT) 30, 1: 3–20. DOI: 10.1504/IJCAT.2007.015693

Sorli M, Stokic D, Mendikoa I, Armijo A (2007b) Advanced IC tools for maximising virtual team creativity and Innovation in Manufacturing environments. INDIN07, 5th International Conference on Industrial Informatics. July, Vienna

Spear S and Bowen H K (1999) Decoding the DNA of the Toyota production system. Harvard Business Rev, Vol. 77 No.5: 96–106

Stamatis DH (2003) Failure Mode and Effect Analysis: FMEA from Theory to Execution. ASQ Quality Press

Sternberg RJ (1999) Handbook of Creativity. Cambridge University Press

Stim R (2006) Patent, Copyright & Trademark. Nolo: 388

Stokic D (2006a) Towards Activity-oriented Collaborative Environments in Manufacturing Industry, Presentation to Collaborative Working Environments FP7 Consultation Workshops on 16–17 March, Brussels

Stokic D (2006b) A New Collaborative Working Environment for Concurrent Engineering in Manufacturing Industry. CE 2006, 13th ISPE International Conference on Concurrent Engineering, Jean-les-Pins, France

Stokic D (2007) Collaborative Working Environment for Innovation in Manufacturing Industry. 13th International Conference on Concurrent Enterprising, ICE 2007

Stokic D, Kirchhoff U, Cordes J, Elfving A, Putz A (1994) Definition of reference architecture for control systems of mobile robots. The 10th ISPE/IFAC International Conference on CAD/CAM, Robotics and Factories of the Future, Ottawa August

Stokic D, Kirchhoff U and Sundmaeker H (2006) Ambient Intelligence in Manufacturing Industry: Control System Point of View. The 8th IASTED Conference on Control and Applications, CA 2006, Montreal, May

Stokic D, Correia A and Grama C (2007) Tools for Designing Collaborative Working Environments in Manufacturing Industry. Proc of the Fourteenth ISPE International Conference on Concurrent Engineering, CE2007

Strohmaier M, Yu E, Horkoff J, Aranda J and Easterbrook SM (2007) Analyzing Knowledge Transfer Effectiveness – An Agent-Oriented Approach. Proc of the 40th Hawaii International Conference on System Sciences (HICSS-40 2007), January 3–9, IEEE Computer Society, Hawaii, USA

Suh N (1990) The Principles of Design, Oxford University Press

Suh N (2001) Axiomatic Design: Advances and Applications, Oxford University Press

Suh N (2005) Complexity: Theory and Applications, Oxford University Press

Taguchi G (1987) System of experimental design. Vols. I and II. UNIPUB, New York

Taguchi G and Chowdhury S (1999) Robust Engineering: Learn how to boost quality while reducing costs & time to market. McGraw Hill Professional, London

Takeda H, Sasaki H, Nomaguchi Y, Yoshioka M, Shimomura Y and Tomiyama T (2003) Universal abduction studio – proposal of a design support environment for creative thinking in design. Proc of ICED'03, 14th International Conference on Engineering Design, Stockholm

Tang P and Molas-Gallart J (2004) Securing Intellectual Property in Collaborative Environments: a Guide. SPRU University of Sussex

Taylor FW (1911) The Principles of Scientific Management. NY Harper Bros

Taylor JE, Levitt and Raymond EA (2004) New Model for Systemic Innovation Diffusion in Project-based Industries. CIFE Working Paper #WP086; Stanford University http://cife.stanford.edu/online.publications/WP086.pdf

Teece DJ (1986) Profiting from technological innovation: Implications for integration, collaboration, licensing and public policy. Elsevier Science Publishers B.V., North-Holland

Teece DJ (1996) Firm Organization, Industrial Structure, and Technological Innovation. J Econ Behav and Organ 31

Terninko J (1996) Step by Step QFD: Customer Driven Product Design. GOAL/QPC. Boston

Thie M and Stokic D (2001) Corporate Knowledge Management in Agile Manufacturing. In the book: Agile Manufacturing, 21st Century Manufacturing Strategy, Elsevier, Oxford

Trickovic I (2006) WS-BPEL 2.0 – Update and Outlook. Objekt Orientiertes Programmieren OOP 2006

Turban E, Wetherbe J and McLean E (2006) Information Technology for Management. 3rd Edition. Lightning Source Inc

Tushman ML and Anderson P (1986) Technological discontinuities and organisational environments. Adminis Sci Quart 31: 439–465.

Tushman ML and O'Reilly, CA (1997) Winning through Innovation. Harvard Business School Press

Ullman DG (2002) Toward the ideal mechanical engineering design support system. Research in engineering design. 13: 55–64.

Ulrich K and Eppinger SD (1995) Product Design and Development. McGraw-Hill Inc
Unit F4 (2005) Collaboration@Work: The 2005 report on new working environments and practices. New Working Environments, Directorate-General Information Society and Media, Office for Official Publications of the European Communities, Luxembourg
Urošević L and Stokić D (2007) ELV Recycling SME Networking supported by Digital enterprise technologies. IFAC Workshop on Manufacturing Modelling, Management and Control 2007, Budapest, Hungary, 14–16. November
Uschold M (2000) Creating, Integrating and Maintaining Local and Global Ontologies. ECAI'00, Berlin, Germany
Uschold M, King M, Moralee S and Zorgios Y (1998) The Enterprise Ontology. Knowl Eng Rev 13, Special Issue on Putting Ontologies to Use. Artificial Intelligence Applications Institute, University of Edinburgh, UK
Utterback JM (1994) Mastering the dynamics of innovation: How companies can seize opportunities in the face of technological change. Boston, MA: Harvard Business School Press
Van der Meer H and Wiles A (2006) ETSI White Paper No. 3; Achieving Technical Interoperability – the ETSI Approach. Sophia Antipolis, October.
Verma E, Chilakapati R and Blanchard B (1996) Quality function deployment (QFD): integration of logistics requirements into mainstream system design. Virginia Polytechnic State University, Department of Industrial and Systems Engineering, Systems Engineering Design Laboratory (SEDL) Department Paper http://www.stevens-tech.edu/sse/fileadmin/sse/625/Appendix_E_QUALITY_FUNCTION_DEPLOYMENT.pdf
Visser U, Stuckenschmidt H, Schlieder C, Wache H and Timm I J , (2002).Terminology for the Management of Distributed Information Resources. Künstliche Intelligenz (KI), Special Issue Knowl Managem, 16(1): 31–34.
Von Hippel E and Von Krogh G (2003) Open source software and the "private/collective" innovation model: Issues for organization science. Organ Sci, 14/2: 209–223.
W3C (2002) Requirements for a Web Ontology Language. http://w3.org/
Wache H (2003). Semantische Mediation für heterogene Informationsquellen. Thesis, Akademische Verlagsgesellschaft Aka GmbH, Dissertationen zur Künstlichen Intelligenz, Berlin, Germany
Wache H, Voegele T, Visser U, Stuckenschmidt H, Schuster G, Neumann H and Huebner S (2001) Ontology-Based Integration of Information – A Survey of Existing Approaches. IJCAI-01 Workshop: Ontologies and Information Sharing, Seattle, Washington, USA
Ward A (2007) Lean Product and Process Development. Lean Enterprise Institute
Ward JK, Liker JJ, Cristiano JJ and Sobek DK. (1995) The second Toyota paradox: How delaying decisions can make better cars faster. Sloan Manag Rev 36/3: 43–61.
Wasson B and Mørch A (2000) Identifying collaboration patterns in collaborative telelearning scenarios. Educational Technology and Society 3, p 3
Watanabe K (2007) Lessons from Toyota's long drive. Harvard Business Rev 85: 7–8
Weber C and Deubel T (2003) New theory-based concepts for PDM and PLM. Proc of ICED'03, 14th International Conference on Engineering Design, Stockholm
WECIDM (2004) Web-enabled Collaboration in Intelligence Design and Manufacture. Asia IT&C ASI/B7-301/3152-99/72553
Wheeler D and Chambers D (1986) Understanding Statistical Process Control. SPC Press
Wille R (1982) Restructuring lattice theory: an approach based on hierarchies of concepts. In Rival I. (ed.): *Ordered sets*. Reidel, Dordrecht-Boston, 445–470
Womack J, Jones D and Roos D (1991) The Machine that Changed the World: The Story of Lean Production. New York Harper Perennial
Yang K and Zhang H (2000) Compatibility Analysis and Case Studies of Axiomatic Design and TRIZ. TRIZ-J September 2000

Yücel G and Aktas E (2008) An Evaluation methodology for ergonomic design of electronic consumer products based on fuzzy axiomatic design. J of Mult-Valued Logic & Soft Comp. 14: 475–493

Zink KJ (1997) Successful TQM: Inside stories from European quality award winners. Gower Publishing Limited

Zlotin A and Zusman A (1999) Managing Innovation Knowledge. TRIZ J February, Southfield, Michigan USA

Zlotin B, Zusman A and Smith L (2002) Futuring the Next Industrial Revolution. ASQ's annual Quality Congress. January. USA http://findarticles.com/p/articles/mi_hb6101/is_200201/ai_n24210451

Further Reading

Crosby PB (1979) Quality is free: the art of making quality certain. McGraw-Hill, New York

CWA3 – CEN Workshop Agreement (2004) European eConstruction Ontology (EeO). CEN/ISSS eConstruction Workshop, Brussels.

FIPA (2001) Ontology Service Specification XC00086D http://www.fipa.org/specs/fipa00086/XC00086D.pdf

Fujita S (1992) Leading an industry by Quality with a warm heart. JUSE, Societas Qualitatis, 6/3 July-August

Galgano, A (1993) Calidad Total. Clave estratégica para la competitividad de la empresa. Díaz de Santos, Spain

Ghoshal S (1997) How the World's leading companies are changing. Management Centre Europe

GOAL/QPC (1990) The Journey. A Traveller's Guide to Total Quality Management. GOAL/QPC, Boston USA

Levy A, Rajaraman A and Ordille K (1996) Querying Heterogeneous Information Sources Using Source Descriptions. Proc of the 22nd VLDB Conference

Lieberman H (1995) Letizia: An Agent that Assists Web Browsing. Proc of the 14th International Joint Conference on Artificial Intelligence, 924–929 Mellish C (ed) Montréal, Canada

Luke S, Spector L, Rager D and Hendler J (1997) Ontology-based web agents. Proc of the First International Conference on Autonomous Agents

OASIS (2008) Open CSA – Service Component Architecture (SCA) Homepage. http://www.oasis-opencsa.org/sca. (last accessed October, 22nd, 2008).

Quemada J, Miguel T, Pavon S, Huecas G, Robles T, Salvachue J, Acosta D, Sirvent V, Escribano F and Sedano J (2005) Isabel: An Application for real time Collaboration with a flexible Floor Control, Proc of CollaborateCom 2005, San Jose, 2005

Trott P (1998) Innovation Management and New Product Development. Pitman

Index

K

L

M

N

O

P

Q

R

S

MIX
Papier aus verantwortungsvollen Quellen
Paper from responsible sources
FSC® C105338

If you have any concerns about our products,
you can contact us on
ProductSafety@springernature.com

In case Publisher is established outside the EU,
the EU authorized representative is:
Springer Nature Customer Service Center GmbH
Europaplatz 3, 69115 Heidelberg, Germany

Printed by Libri Plureos GmbH
in Hamburg, Germany